ALPINE PHYSICS

Science in the Mountain Environment

ALPINE PHYSICS

Science in the Mountain Environment

Valerio Faraoni

Bishop's University, Canada

World Scientific

NEW JERSEY · LONDON · SINGAPORE · BEIJING · SHANGHAI · HONG KONG · TAIPEI · CHENNAI · TOKYO

Published by

World Scientific Publishing Co. Pte. Ltd.

5 Toh Tuck Link, Singapore 596224

USA office: 27 Warren Street, Suite 401-402, Hackensack, NJ 07601

UK office: 57 Shelton Street, Covent Garden, London WC2H 9HE

British Library Cataloguing-in-Publication Data
A catalogue record for this book is available from the British Library.

ALPINE PHYSICS
Science in the Mountain Environment

ISBN 978-981-3274-20-4

For any available supplementary material, please visit
https://www.worldscientific.com/worldscibooks/10.1142/11097#t=suppl

Desk Editor: Nur Syarfeena Binte Mohd Fauzi

Typeset by Stallion Press
Email: enquiries@stallionpress.com

To Matteo Vianini

Preface

According to the United Nations organization, mountains cover one quarter of the land of this planet, are home to a billion people, and provide over 70% of the freshwater available on land. For these reasons, December 11 has been designated International Mountain Day. Mountains are the essence of mountain cultures and the passion of mountain climbers. Mountains are hence important and, when trying to understand the alpine environment, one realizes that there are both very simple laws of physics at work in the mountains as well as some more complicated ones. Although these laws are basic and fairly easy to understand, their importance is not fully appreciated as easily, though one should not forget them. A little reasoning shows why. These laws have numerous and very important consequences for those who live or travel in mountain terrain. One should be aware of them at all times when one is out there, and must understand the environment and its mutable conditions in order to make critical decisions on which a climber's life — and those of the members of a climbing party — may depend. In more ordinary situations when no drama is involved, seeing and understanding these physical laws in action will just add to the appreciation of the mountain environment. For example, you may be lying down on a well deserved rest after a climb, enjoying the sun and wondering about what lies in front of you. The mountains are steep terrain, often vertical or even overhanging, and on this terrain gravity rules. The dictatorship of gravity is felt by a body hiking up a steep trail or climbing a steep face or a slippery slope on snow or ice. Gravity pulls on the snow which may come down on us as an avalanche when the conditions are right. Gravity pulls on rocks and chunks of ice that have the bad habit of detaching, threatening, and falling on passer-bys. Gravity acts on water that flows as impetuous mountain streams, on slides which erase significant portions of the face of a mountain and obliterate classic climbing routes with or without climbers on them, and on boulders which mysteriously dislodge and roll down a slope. Gravity acts on water rushing down a wall in a thunderstorm, sweeping gullies and small channels and carrying rocks and debris with it, and it acts on snow sloughing off a cliff. Gravity pulls down cornices breaking on mountain ridges and it destroys dicey snow bridges across crevasses. But there isn't only gravity at play: there is a lot of other physics involved in these phenomena and their detailed modelling could keep scientists busy for a long time (in fact, it often does). Not that the mountaineer cares much about physical modelling and its mathematical nuances when caught in dire situations, or panting uphill on a steep slope under a big load, but a little understanding and physical sense can contribute to avoiding bad situations in the first place. An experienced mountaineer has seen many things happening in the hills and has built an understanding of what is going on and the ability to predict what will happen in a certain place under certain conditions, plus the ability to forecast these conditions. Over the years, he or she has built a general sense of the mountain environment and a large part of this understanding is just what physicists call physical sense, applied to the specific mountain environment. There is an innate pleasure in understanding the science related to the mountains and one of the main goals of mountaineering, after all, is maximizing the amount of fun per unit time. It is great to sit on top of a peak and visualizing in thirty seconds the erosion process by glaciers that gave that valley in front of us its U-shape over a few tens of thousands of years. And then we realize that the stripes on those smooth granite slabs on which we are resting were caused by huge ice masses grinding them or, more precisely, by boulders stuck in the ice and acting as a giant file.

Although gravity is dominant, there is much more physics involved in mountaineering, including mechanics, meteorology, hygrometry, fluid mechanics, thermal and statistical physics, electromagnetism, and non-linear physics. There is an enormous range of applications of physics to the natural phenomena encountered in mountaineering: electromagnetism becomes obvious when we are late descending from a climb and we get caught by the afternoon thunderstorm on a summit or along a ridge. Then we wonder what is the preferred path of a lightning bolt, what is the probability of being hit in that particular place, and we would like to know more about ground currents when we stop in a "best place" in thunderstorm position.

The microscopic cohesion forces between snowflakes of a settling snowpack which metamorphose over time are crucial when trying to assess whether to ski a suspicious slope or to back off. The importance of phase changes is perceived when they dislodge rocks above us as the temperature oscillates around zero degrees Celsius. The constant pull of gravity cannot be forgotten when contemplating the effects of erosion, the great destroyer of mountains, for amusement or for trail maintenance. Not that one starts making physics calculations with paper and pencil or that one runs a Monte Carlo computer simulation when caught in a tricky situation, but a little reasoning can help in understanding those situations and making sense of things later, when there is more time to think and to wonder how the events unfolded.

Historically, many of the pioneers of mountaineering in the European Alps and in other areas of the world were trained scientists and, in those early days, the scientific spirit was mixed with the spirit of adventure and discovery [Newby (1977); Davis (2012); Pole (2005, 1991); Worster (2008); Richardi (2008)]. These pioneers often carried heavy instruments to perform scientific measurements on top of the peaks climbed, to determine their elevation, to sketch geographic maps of the area, or for their atmospheric studies. The great Viennese mountaineer Paul Grohmann (1838–1908) who made many first ascents in the Dolomites of Italy (then a part of the Austrian-Hungarian empire) tells tales of how hard it was to climb technical peaks in the Dolomites without breaking his precious barometer, and how he would not let his guides carry it but he would transport the fragile instrument himself [Grohmann (1877)]. The Irish physicist and pioneer mountaineer John Tyndall (1820–1893) mixed physics research and science popularization with mountaineering at a time when both were becoming popular [Reidy (2010)]. It is widely recognized that the development of science and of a scientific attitude pushed the exploration of previously unknown mountains and helped the development of mountaineering in its early stages.

This book examines rather obvious applications of physics to the mountain environment and is written for mountaineers, lovers of mountains, or curious people who wonder why or how this or that natural phenomenon happens, what causes it, and what its scientific explanation may be. This book is not a systematic analysis of the mountain environment and the physics discussion does not go nearly as deep as it could. Indeed, the science of mountain phenomena can quickly become very involved and the research is often open-ended. Sometimes we do not have a clear answer to a scientific question, and that's what makes the subject attractive to researchers in the first place. However, this book is written mostly for non-scientists. While a physics textbook would organize the topics according to the branch of physics to which they belong, I prefer to group the material in categories more relevant for the mountain climber.

It is said that a picture is worth a thousand words. In turn, physicists say that an equation is worth a thousand pictures. Therefore, we will see some formulae and some more technical explanation but, most of the times, these can be skipped without jeopardizing the understanding of the rest. Physical sense is often just plain common sense and mountaineers relate to that. This physical sense, augmented by a little knowledge, is a tool to further the understanding of the mountain environment. This book is not a mountaineering guide and, of course, mountaineering and climbing are not learned from books. As stated on the disclaimers accompanying all the pieces of gear that we buy, there is no substitute for experience, a statement which is reinforced by physicists since physics is an experimental science. This book is not conceived as a preparation for the outdoors, but it is more of a reading after one comes back tired from a mountain trip and has no immediate urge to go back out again, or when the weather is bad and we are stuck at home. A friend of mine even carries such books on multi-day trips to read in the tent in case of rain, which happens all too often in the mountains of coastal British Columbia where he climbs. Some of this book may, after all, turn out to be of some practical use — other than starting a campfire, and I will be happy if it does, but it should not be taken as a practical manual.

Physics and fundamental science in general are seen at work everywhere in nature, but in the mountains we see processes that are much faster or more dramatic than in other situations because of the large gradients involved. It helps to be prepared for the situations that mountains create, to understand them, and to enjoy them as safely as possible.

Those who know some physics will only whet their appetite by reading this book and will probably be mad at its author for not going into some detail. This book provides only glimpses of selected topics in physics and science. It is not possible to make justice to the complexity and variety of physical phenomena encountered in alpine terrain in a short introductory exposition. To be forgiven by those readers, we will see various references that point them to more satisfactory discussions. Finally, the International System of Units (SI system) is used, and the names of my friends appearing in some of the stories of this book have been changed.

Valerio Faraoni

An important note to readers

Mountaineering and rock climbing are dangerous activities which involve the risk of serious injury and death. Before engaging in these activities you should take proper instruction from experts and acquire the necessary competence and fitness. Climbing and mountaineering are not learned from books and this is not a climbing instruction book anyway. When mountaineering, climbing, hiking, skiing, biking, or canyoning, you are solely responsible for your own safety and for that of your climbing party, and for understanding and judging the environment around you. It is obvious that the author and the publisher of this book cannot in any way be responsible for accidents resulting from the climbing or mountaineering activities described in this book. You always climb or hike at your own risk.

List of Notations

m, M	mass
V	volume
A	area
g	acceleration of gravity
G	gravitational constant
h, z	elevation or thickness
r, R	radius
t	time
$\vec{x}$	position
$\vec{v}$	velocity
$\vec{a}$	acceleration
$\vec{F}$	force
$\vec{N}$	normal force or torque
P	pressure
ρ	mass density
τ, σ	stress
E	energy
$\vec{p}$	momentum
T	kinetic energy
V	potential energy
L	Lagrangian
Q	heat energy
T	temperature

L_f	latent heat of fusion
L_v	latent heat of vaporization
γ	surface tension
B	bulk modulus
f	shape factor (of valley glaciers)
α	slope angle
ν	frequency of electromagnetic waves
λ	wavelength
I	intensity (of radiation)
ϵ	molar extinction coefficient
μ_s	coefficient of static friction
μ_k	coefficient of kinetic friction
q	electric charge
$\vec{E}$	electric field
a	scale factor of the universe
H	Hubble parameter
c	speed of light
R	universal gas constant
K	Boltzmann constant
N_A	Avogadro's number
D	diffusion coefficient
h	Planck constant
$\hbar = \dfrac{h}{2\pi}$	reduced Planck constant
$\vec{\nabla}$	gradient
∇^2	Laplacian
$\equiv$	equal by definition
$\simeq$	approximately equal
$\propto$	proportional
$\perp$	perpendicular
$\parallel$	parallel
$\mathrm{Tr}(\hat{\mathbf{A}})$	trace of the matrix $\hat{\mathbf{A}}$

Acknowledgments

I am grateful to all the people with whom I climbed and hiked over the past thirty years. They are too many to list, but they are all remembered. I am especially grateful to the friends who introduced me to mountaineering, rock climbing, skiing, and canyoning in Italy and in Canada. A special thank you goes to my physics collaborators, including my students and former students, who share the passion for the mountains and with whom I had many physics discussions while climbing one. Many thanks to Louine Niwa for her help with the artwork. Finally, I am grateful to Dr. Don Mak, World Scientific Editor, for the encouragement in bringing this book project to completion.

Contents

List of Table

List of Figures

Chapter 1

Mountains high or low, hard or soft

1.1 Introduction

Fortunately there are mountains, plenty in some geographical areas and none in others. Mountains come in various shapes, big and small, in ranges or as isolated volcanoes sticking out of the plains; mountains old or young, glaciated or rocky, and their walls, peaks, and ridges can be made of igneous, sedimentary, or metamorphic rocks which determine the slopes and the shape of the landscape. Mountains host glaciers and permanent snowfields, wide U-shaped valleys carved by glaciers, and narrow gorges through which impetuous streams run (Figs. 1.1 and 1.2).

These features give rise to a variety of physical phenomena, some of which are going to be discussed. Perhaps the most basic characteristics of mountains are hiding in plain sight and are easily missed, for example the simple statistics that no mountain on Earth is taller than about nine kilometers, or that there is a limiting thickness not only for alpine glaciers but also for the huge polar glaciers which are a few kilometers thick. There are physical reasons why things are like that. Geological forces are active building mountains and other forces destroy them simultaneously, but they can ultimately be reduced to thermal processes inside the Earth or to mechanical processes dominated by gravity on the outside. Basic concepts of physics explain features of these mountains and mathematics has something to say about them. Let us look at some of these concepts applied to the mountain environment.

Fig. 1.1   The Miage torrent rages through the V-shaped Gorge de la Gruvaz (Val Montjoie, Saint Gervais les Bains, France).

1.2   How high can a mountain be?

How high can a mountain be? The summit of the highest mountain on Earth, the legendary Mt. Everest, stands 8848 meters above sea level. If one measures the elevation starting from the bottom of the ocean,

Fig. 1.2   Canyoning in a limestone gorge carved by running water (Val Clusa, Italy).

Mauna Kea (4,207 meters above sea level and over 10,000 meters from the floor of the Pacific Ocean) is definitely taller than Mount Everest. Mountains are not arbitrarily high and 10 km is still a very small fraction of the radius of this planet (which is approximately 6370 km) so, relatively speaking, mountains are very small warts on the surface of our planet. In comparison, the corrugations on the peel of an orange are larger. What limits the height of a mountain? Clearly, it must be physics together with the basic properties of the materials that mountains are made of.

Most geography and geology textbooks give the answer: the maximum height of a mountain on Earth (approximately 10 km) is set by the basic physical fact that an hypothetical mountain higher than this limit would begin to melt at its base, which would flow away, hence mountains higher than about 10 km cannot exist. It is interesting that the same geography textbooks never seem to calculate this 10 km limit, nor do they provide references to support the argument (the nice picture of a mountain that is often included at this point does not do much, in terms of logic, to support or illustrate this physical argument). The calculation is really quite simple and, to the best of our knowledge, the argument was proposed by Victor Weisskopf in the 1970s [Weisskopf (1975)]. Weisskopf (1908–2002) was an Austrian-born, American-naturalized theoretical physicist who worked on building the theory of quantum mechanics with the likes of Werner Heisenberg, Erwin Schrödinger, Wolfgang Pauli, and Niels Bohr and later took up a leading role in the Manhattan project (after World War II, however, he campaigned against nuclear proliferation). True to his Austrian background, Weisskopf must have liked the subject of the maximum height of mountains because he discussed it again eleven years later in the famous pedagogical journal *American Journal of Physics*[1] [Weisskopf (1986a)] (a brief debate also ensued [Landis (1986); Weisskopf (1986b)]) and there are records of a lecture on this subject given by Weisskopf at CERN, the European organization for research in particle physics located in Geneva [CERN (2017)].

Weisskopf's argument goes like this[2]: consider a horizontal parallelepiped (a very crude "mountain") made of a most common mineral, say SiO_2, and think of making this parallelepiped thicker and thicker. This "mountain" reaches its maximum height when a layer of the same material added to its top causes plastic flow at the bottom. Weisskopf's trick to calculate the effect and simplify the problem consists of noting that this process amounts to melting a layer of equal thickness at the base of the "mountain" (Fig. 1.3).

One does not need to actually lower this layer from the top to the bottom: we can just consider that a layer added at the top causes a layer of the same thickness to melt at the bottom and we can *imagine* to move the top layer to the bottom. In fact, it is sufficient to provide only a fraction η of the energy needed to melt this layer, and it is provided by the loss of gravitational potential energy (the energy that a mass has just by virtue of being up high in the presence of

---

[1]Unlike research journals, this is a professional journal completely devoted to improving and simplifying the teaching of university physics.

[2]A similar argument with more emphasis on the tensile properties of materials is given in Scheuer (1981).

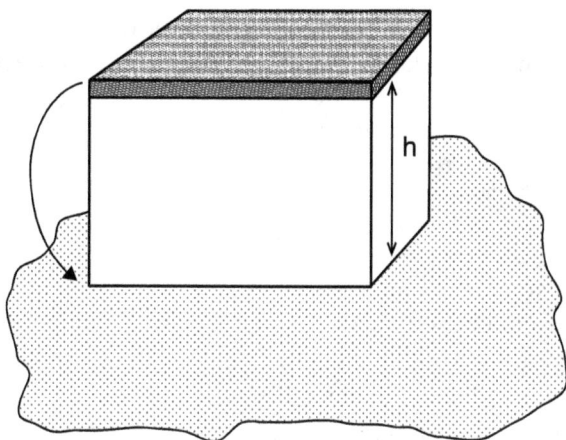

Fig. 1.3 Weisskopf's argument.

gravity). Weisskopf guessed the value of this factor as $\eta \sim 1/3$. So, we equate the gravitational energy lost by lowering the top layer to the base of the "mountain" with one third of the thermal energy necessary to melt a layer of the same thickness. The latter is well known from everyday experience: as is intuitive, to melt a mass m of a material one needs to provide an energy proportional to this mass

$$E = L_f m,$$

where L_f is a constant characteristic of the material and called *latent heat of fusion*.[3] In a student lab the melting energy E is supplied as heat, but here it is provided by the gravitational energy which the layer of material had by virtue of being at the top of the mountain.[4] The gravitational potential energy of a mass m at level h with respect

---

[3]The temperature does not change during the process, if the material is chemically pure and the pressure stays constant. Usually melting happens at atmospheric pressure, so we won't worry about that here.

[4]If you are not happy with this explanation, you can think of the weight of the mountain causing pressure at the bottom, which heats up the material. This process happens when we make a snowball by squeezing hard the snow in our hands, which causes the ice crystals in the snow to melt (and then refreeze and bind together) even at sub-zero temperature.

to a reference level (in this case the bottom of the mountain) is mgh, where $g = 9.8\,\text{m/s}^2$ is the acceleration of gravity. We have then

$$mgh = \eta\, L_f m$$

and, solving for h,

$$h = \frac{\eta\, L_f}{g}. \tag{1.1}$$

The mass m cancels out and disappears from our result, which is good, otherwise we would obtain different numbers for every different layer that we could want to consider, and all those layers would have different masses. The latent heat of fusion of SiO_2 is tabulated as $L_f = 2 \cdot 10^5\,\text{J/kg}$ [Clauser (2011)], which gives

$$h = \frac{2 \cdot 10^5\,\text{J/kg}}{3 \cdot 9.8\,\text{m/s}^2} = 7\,\text{km}.$$

To build our hypothetical "mountain", we have used SiO_2 but we could have used instead basalt or granite, which are two other common rocks. In this case the latent heat of fusion is $L_f = 4.2 \cdot 10^5\,\text{J/kg}$ [Clauser (2011)], which would provide the maximum height of approximately $14\,\text{km}$. The calculation presented is what physicists call an *order of magnitude estimate*. It is not meant to be a precise calculation, which would require a lot more knowledge of physics, detailed modelling, and more mathematical sophistication, but provides only an answer correct within a factor of a few with little effort. Weisskopf's argument identifies the key physical elements of the problem (the strength of the atomic/molecular bonds within the material, which determines the latent heat of fusion, and gravity) and does the job for us.

1.3   Out of this world

We have just seen that the maximum height of a mountain made of a material with latent heat of fusion L_f is

$$h \simeq \frac{L_f}{3g},$$

where g is the surface gravity of the planet. Although we will never get there for a weekend trip or even for an extended climbing expedition, it is inspiring to think of mountains on other planets where the acceleration of gravity g is different from $9.8\,\mathrm{m/s^2}$, the average value on Earth.[5] We refer here to rocky planets like the Earth, for it would not make sense to speak of mountains on a gaseous planet like the gas giants of our Solar System Jupiter and Saturn. Take for example Mars, which has a surface gravity g equal to $0.376\,g_{\mathrm{Earth}}$. By replacing g in our previous equation, we obtain

$$h_{\mathrm{Mars}} \simeq \frac{L_f}{3 \cdot 0.376\,g_{\mathrm{Earth}}} \simeq \frac{h_{\mathrm{Earth}}}{0.376} \simeq 27\,\mathrm{km}$$

using $h_{\mathrm{Earth}} \simeq 10\,\mathrm{km}$. Because gravity is weaker than on Earth, mountains on Mars can be taller than their terrestrial cousins. For example the tallest mountain on Mars, Olympus Mons, a shield volcano, is approximately $22\,\mathrm{km}$ high.

The maximum height of a mountain on a planet is related to another topic which interests astrophysicists. Thinking about it, it is easy to compute the surface gravity of a planet using Newton's theory of gravity if we know its mass M and its radius R (what is not easy is to measure the mass and the radius of distant planets, but that has been done with many years of astronomical studies). The force of gravity acting on a small mass m placed on the surface of a spherical planet (that is, at a distance R from its centre) is mg, where g is its surface gravity. The attraction between the entire planet of mass M and the small mass m is described by Newton's famous law of universal gravitation, which states that two masses M and m at distance r attract with a force of intensity

$$F = \frac{GMm}{r^2},$$

---

[5]We need not limit ourselves to the Solar System: thousands of exoplanets, worlds beyond our Solar System, have been discovered and there can be a large variety of rocky planets with many possible values of the surface gravity g [Jayawardhana (2011)]. Many of these exoplanets may harbour life, and maybe even mountaineers and alpine clubs. The search for, and study of, exoplanets is currently a hot area of research.

where $G = 6.67 \cdot 10^{-11}\,\mathrm{N \cdot m^2/kg^2}$ is the gravitational (or Newton) constant. Newton also told us that a spherical planet behaves as if its mass were concentrated at its centre (the *iron sphere theorem*). Then, for a particle on the surface of a planet, we set the distance r equal to the radius of the planet R and we have

$$F = mg = \frac{GMm}{R^2}.$$

The small mass m cancels out and the value of the surface gravity is the same,

$$g = \frac{GM}{R^2}, \tag{1.2}$$

for *all* masses m. In other words, if no other forces (air friction, *etc.*) are present, all bodies (and all climbers) fall with the same acceleration independent of their mass or internal composition. This key fact, which goes against the erroneous intuition of Aristotle and of the ancient Greek philosophers (who didn't bother experimenting), was established empirically by Galileo Galilei in 1589. It is called *universality of free fall* and is peculiar to gravity. None of the other three fundamental forces of nature (electromagnetic force, strong nuclear force, and weak nuclear force) share this feature. This simple fact is the basis of Einstein's theory of General Relativity which describes gravity in modern physics[6] [Carroll (2004); Wald (1984); Hartle (2003); Misner, Thorne, and Wheeler (1973)]. But let's proceed with our subject. If we substitute in Eq. (1.2) the values of the mass and radius of the Earth $M = 5.978 \cdot 10^{24}$ kg and $R = 6370$ km, we obtain the value of the acceleration of gravity $g_\text{Earth} = 9.8\,\mathrm{m/s^2}$ that we started our discussion with, and that we measure experimentally.

---

[6]Einstein's reasoning is that, if gravity is a universal force which is the same for all falling bodies, it is possible to describe it as a property of space instead of a force. In General Relativity, small masses are not subject to any gravitational force but move freely along trajectories in a space that is curved by other masses [Carroll (2004); Wald (1984); Hartle (2003); Misner, Thorne, and Wheeler (1973)]. Only, being a theory of relativity, it is not only space that is curved but also time. As American physicists J.A. Wheeler put it, "matter tells space how to curve and space tells matter how to fall" [Wheeler and Ford (1995)].

This is all fine for a spherical planet. However, in the space around our planet and around other Solar System planets, we see asteroids and other bodies which have irregular shape, like potatoes. We also see that larger bodies, however, assume spherical shape apart from very minor irregularities on the surface which are the mountains that, on Earth, constitute the delight of mountaineers. These irregularities are minor indeed: the height of mighty Mt. Everest, 8848 m, is a very small fraction of the average radius of the Earth, 6370 km:

$$\frac{8848 \text{ m}}{6.37 \cdot 10^6 \text{ m}} = 0.00139.$$

A celestial body can only have irregular shape if its size is not too large. Mountains on this body (say, an asteroid) can reach heights of the order of the size of the entire body itself (which is another way of saying that it has irregular shape) only if the surface gravity of this body is not too large. When the body is large, its gravity overcomes the rigidity of the material it is made of, large mountains melt away at their base as discussed in the previous section, and the shape of the body becomes spherical. The maximum size that a body can have and still admit fairly large irregularities, like a potato, is known in astrophysics as the *potato radius*. The potato radius separates asteroids from dwarf planets and it can be easily calculated as follows [Lineweaver and Norman (2009); Caplan (2015)]. Our equation (1.1) for the maximum height of a mountain, which we derived in the previous section, gives

$$hg = \eta L_f,$$

where ηL_f, and therefore hg, is a constant characteristic of that material. Now assume that an asteroid is made of the same material as the Earth, which is reasonable as an approximation, then the product hg for this asteroid is the same as that for the Earth, $h_{\text{Earth}} g_{\text{Earth}}$. Assume, further, that the asteroid is reaching the potato radius, then it is assuming spherical shape with a radius $R \sim h$ so that

$$Rg \simeq h_{\text{Earth}} g_{\text{Earth}}.$$

The gravity of the asteroid is approximately $g = GM/R^2$ and, assuming it has homogeneous density ρ, its mass is

$$M = \frac{4\pi}{3} \rho R^3,$$

which gives

$$R \frac{\frac{4\pi G}{3} \rho R^3}{R^2} = h_{\text{Earth}} \, g_{\text{Earth}}$$

and, finally,

$$R \simeq \sqrt{\frac{3 h_{\text{Earth}} \, g_{\text{Earth}}}{4\pi G \rho}}.$$

Using as an indicative value for the asteroid density $\rho = 5.5 \cdot 10^3 \, \text{kg/m}^3$, which is the average density of the Earth, $h_{\text{Earth}} \sim 10 \, \text{km}$, $g_{\text{Earth}} = 9.8 \, \text{m/s}^2$, one obtains $R \sim 250 \, \text{km}$, which matches the observed values for asteroids. Mountains are interesting also outside of this world.

1.4 Fallen from the sky

Mountains are remarkable, as we know, but those on Iapetus are really remarkable. Iapetus is a moon of Saturn and it has an equatorial ridge system with peaks that reach the height of 20 km and is about 70 km broad. Ridges have slopes reaching $40° - 45°$. These features have got scientists thinking hard. Could this mountain ridge have originated because of tectonic movements, as is normal on Earth? Could its origin be volcanic? The answer is not clear at the moment but the favourite theory is that of an exogenic origin which, translated in mundane language, means that these mountains literally fell from the sky. The people proposing this model are reputable scientists (which probably means they are from mildly eccentric to real nuts, but not outright crazy), so read a little further before you put down this book. Saturn is famous for its beautiful rings and planets and their moons are expected to have some rings, too. A ring of rocks and dust could have formed early around Iapetus and then could have been destroyed by cosmic impacts. This is not fiction, for there is plenty of evidence of cosmic impacts. Craters on the Moon and on celestial bodies without an atmosphere and an hydrosphere to erase them are proof. We even have meteor craters on Earth, although much less, and much less evident because of erosion. Another piece of evidence for the abundance of cosmic collisions was the spectacular impact of comet P/Shoemaker-Levy 9 on Jupiter observed in July 1994. The ring around Iapetus would have been located in a plane around the equator, which would explain in a very simple way the location of the equatorial mountain ridge observed. Moreover, if the working hypothesis is correct, the debris should have deposited along steep slopes near the angle of repose (see Sec. 2.3). Indeed, recent studies of the topography of the Iapetus ridge [Lopez Garcia *et al.* (2014)] show that the slopes at various locations along the ridge are close to the angle of repose for rounded grains (approximately 25°) or for snow mixed with hail/graupel (approximately 45°). As normal, scientists try to poke holes into each other's theories and it is only after intense scrutiny that a scientific theory can stand. This process is normal and is an essential part of the scientific method, contrary to pseudo-science in which a favourite "result" is instead assumed from the outset and a selection of "facts" or reasonings is assembled to support the pseudo-theory, while all evidence to the contrary is conveniently ignored. Instead, science always attempts to falsify a theory and to establish its limits of application, to perform experiments or observations which support or destroy it, and to examine them objectively. Results are published only after peer review and are then exposed to the experimental and theoretical scrutiny of peers who attempt to reproduce the results and then confirm or dispute them.

To be consistent with the hypothesis that the mountains on Iapetus fell from the sky, the ridge on this moon should be uniform as the longitude changes, and its transverse cross-section should show a single prominent peak. This is not the case, as the ridge is broken into many peaks and the cross-sectional morphology is more complicated, with several types of peaks observed: triangular, trapezoidal, crowned, twinned, saddled, and irregular [Lopez Garcia *et al.* (2014)]. However, things are not so simple and much can happen over an astronomical time scale. Given an impact with a cosmic object that destroys a ring and creates the equatorial ridge, it is very likely that further impacts would have occurred and some could have modified the morphology of the equatorial mountain ridge. It is even possible that a smaller satellite of Iapetus could have formed and been destroyed by tidal forces, with the resulting debris falling onto the equator of Iapetus [Lopez Garcia *et al.* (2014)].

Another satellite of Saturn, Rhea, is believed to have a similar equatorial ridge [Schenk *et al.* (2011)], so it is possible that events like the one hypothesized here are indeed common in the Solar System. Future studies and detailed numerical modelling will hopefully shed light on this fascinating, albeit exotic, aspect of mountain science. For the moment the argument that, outside of Earth, some mountains can literally fall from the sky is rather strong.

1.5   All the peaks, passes, and valleys of this planet

Suppose that somebody has counted all the peaks and passes on Earth. Would it then be possible to know right away the number of valleys on the planet without counting them? Intuitively, one must cross a pass to go from one valley to another; there must be a valley on each side of a pass, or it wouldn't be a pass, and there are peaks on both sides of a pass, or it wouldn't be a pass, so clearly there is some relation between number of valleys, number of passes, and number of peaks. But is this observation sufficient to answer the question? In fact it is, according to a branch of mathematics called differential topology. A theorem due to the eighteenth century French mathematician Henri Poincaré and to the twentieth century German mathematician Heinz Hopf establishes the relation. The *Poincaré–Hopf index theorem* states that the number of peaks n_{peaks} plus the number of valleys n_{valleys} minus the number of passes n_{passes} on a sphere (and the Earth is a sphere from the point of view of topology[7])

---

[7] *Topology*, a branch of mathematics, concerns itself with the shape of objects. To give an idea, a doughnut with its central hole and a teacup with a handle are two surfaces topologically equivalent to each other, but topologically distinct from a sphere, which has no holes nor handles [Singer and Thorpe (1967)].

equals two:

$$n_{\text{peaks}} + n_{\text{valleys}} - n_{\text{passes}} = 2.$$

This equation answers the question posed and establishes a precise (linear) relation between the numbers of peaks, valleys, and passes. Suppose that we want to know how many passes there are, given the number of valleys and the number of peaks. The number of passes grows linearly with the other two variables n_{valleys} and n_{peaks}. For example, if we have a ridge separating two valleys, as in Fig. 2.3, one can add peaks to the ridge and the number of passes, which always separate two adjacent peaks, will grow with n_{peaks} .

Now the more technical part: the theorem states that the sum of the indices of the isolated zeros of a vector field on a surface of genus γ is $2 - 2\gamma$. One can regard the elevation h above sea level as a regular function of two coordinates (two angles, for example latitude and longitude) on the surface of an otherwise spherical Earth (okay, it is an idealization[8] but it works). The gradient $\vec{\nabla} h$ of the elevation above sea level is the vector field to be used in the theorem, since it is a vector (pointing in the direction of maximum increase of elevation and perpendicular to contour levels of constant elevation) which depends on the location. The gradient $\vec{\nabla} h$ of h vanishes at maxima, minima, and saddle points of the elevation (thought as a function of the position on the sphere), which are assumed to be isolated. The index of a maximum of h (*i.e.*, a peak) is $+1$, the index of a minimum of h (*i.e.*, a valley bottom) is $+1$, the index of a saddle point (*i.e.*, a pass) is -1, while a sphere has genus $\gamma = 0$, which gives the result quoted.[9] The more precise formulation of the theorem is that the sum of all the indices x^i of the isolated zeros (labelled by the letter i) of the vector field equals the genus (or "Euler characteristic") of the manifold on which this vector field is defined (in this case the

---

[8]Idealization is the essence of mathematical modelling in science, and especially in environmental science (*e.g.*, Harte (1988, 2001)).

[9]For comparison, a doughnut, which is topologically distinct from a sphere, has genus 1. Remember that the doughnut has the same topology of a teacup with one handle. The genus of a surface can be described as the number of handles it has [Singer and Thorpe (1967)].

surface of the Earth), which must satisfy some mild assumptions, or

$$\sum_i x^i = \gamma.$$

The mathematical significance of the theorem is that the peaks, passes, and valleys are purely local concepts described by the elevation of these points and of some small areas around them, while the genus, known from entirely different considerations, is a *global* characteristic of the entire Earth, that is, a topological feature. The theorem links two apparently unrelated areas of mathematics, analysis (local) and geometry (global). In other words, the theorem means that the total numbers of peaks, valleys, and passes somehow "know" that they live on a sphere while a single peak or pass doesn't. Magic! Of course, this mathematical theorem has no practical relevance for a mountaineer but think about it anyway when you contemplate many distant peaks, passes, and valleys from a high summit.

1.6   The rock cycle

Geologists classify rocks according to their origin as *igneous, sedimentary*, or *metamorphic*. Rock climbers and mountaineers may not know geology and mineralogy, but they instinctively know a lot about rocks and they appreciate their differences, textures, and the variety of holds that they can offer or deny to them.

Igneous rocks are further classified as *plutonic* and *volcanic*. Plutonic or intrusive rocks are formed when magma cools slowly and hardens below the surface of the Earth and is later exposed (the name comes from Pluto, the Roman god of the underworld). Plutonic rocks include granite, diorite and gabbro, plus other rocks like syenite, granodiorite, monzonite (known, for example, to rock climbers playing in Joshua Tree, California), piroxenite, and peridotite.

Volcanic or extrusive rocks, as the name says, are associated with volcanism and include basalt, rhyolite, andesite, and also other rock types like tuff, obsidian, and pumice. Basalt, well known to rock climbers in some areas, is the most common volcanic rock on Earth. It forms ocean floors and up to two thirds of the Earth's crust are made of basalt. And this statement does not hold only on Earth: basalt forms the *maria* (Latin for "seas") seen as dark patches on the moon. Andesite, as the name says, is common in the Andes of South America, which were created by andesitic lava flows.

The slow cooling of plutonic rocks is responsible for their highly ordered crystalline structure: by contrast obsidian, another volcanic rock, solidifies so fast that crystals are not present in it and it is called a "volcanic glass". Like ordinary glass, it has no crystalline structure and is more similar to an extremely viscous fluid. It tends to fracture with very sharp edges (hence its early use to manufacture arrowheads and knives), does not form walls, offers no handholds and footholds (which would be too sharp anyway), and is of no interest to rock climbers but is still fascinating to look at. Plutonic rocks, instead, generally have large and visible crystals (Fig. 1.4).

Granite, which makes magnificent mountains and provides exceptional climbs such as those of Yosemite National Park in California or of the Mont Blanc needles in France and Italy, is an igneous plutonic rock. Granite typically contains large crystals and is coarse-grained (Fig. 1.4).

The more coarse-grained granite makes for excellent friction climbing and also grinds the skin of climbers slipping on those long runout slabs like a grater (we wear long pants and long sleeves on slabs and we tape our hands for granite cracks, or we suffer). There is also finer-grained granite which does not provide as much friction but nevertheless attracts at least local rock climbers. Because of its compactness, granite tends to fracture in blocks or to offer cracks. For this reason, granite climbing usually means crack climbing, friction climbing, or the occasional use of dykes. These features are intrusions of quartz or other rock very different from the surrounding rock in color and consistence, and are discordant with the surrounding layers (Fig. 1.5). They formed when minerals deposited into fissures of the rock and, although sharp, they can provide a much appreciated break when leading on a granite slab runout.[10]

---

[10]There are also clastic dykes, formed when sedimentary rock fills an existing crack.

Fig. 1.4   Large visible crystals in granite.

Basalt, another volcanic rock, is formed when magma cools rather rapidly at the surface and only small crystals form. As a general rule, crystals need time to grow in a long, slow cooling process and volcanic rocks which cool fast have no crystals or only small crystals.

Porphyry contains both large and small crystals, of very different size. The larger crystals form underground and are later taken to the surface by lava, where smaller crystals form and surround the larger ones.

Fig. 1.5   A dyke in granodiorite (Tuca Blanca, Maladeta range, Spain).

Sedimentary rocks are formed by the erosion of pre-existing rocks, which are transported, deposited and, together with sand, plants, shells, and pebbles, accumulate and consolidate forming layers which are gradually compacted into new rock and often tilted by the movement of parts of the Earth's crust. They are then exposed to air by erosion. Sedimentary rocks include limestone, dolomite, sandstone, conglomerates, mudstone, and shale. Sedimentary rocks are usually layered and the layers are often curved, bent, twisted, and inclined with respect to the horizontal. Sometimes the layers have different colors. Sedimentary rocks are softer than igneous rocks. The Dolomites of Italy and the limestone Kaisergebirge of Tyrol comprise many spectacular ranges and provide endless rock climbs on towers, faces, and chimneys, in addition to a variety of hiking trails and, because of their beauty, alpine huts and bars from which to enjoy the superb scenery. Dolomite rock takes its name from the French mineralogist and geologist Deodat Guy Tancrede Gratet de Dolomieu (1750–1801), the discoverer of its chemical composition. Dolomite is a double carbonate of calcium and magnesium, $CaMg(CO_3)_2$, which can be eroded to form many small holes, pockets, and holds of various shapes, very attractive to rock climbers.

Sedimentary rocks often host fossils embedded in them, which by contrast are extremely rare in igneous or metamorphic rocks. In addition to informing us about ancient life forms, fossils are useful because they indicate the age of the sedimentary rock and provide information about the ancient climate and the environment in which they formed.

The commercial and geological names of rocks do not match. For example, sedimentary rocks are called "marbles" in the stone industry, while this name is reserved to metamorphic rocks by geologists. Ammonites are found in rocks called "marble" in quarries in many regions of Europe, but this is really sedimentary rock. For example, Rosso Verona (Verona red) is a reddish sedimentary rock always called a "marble" in the industry. It is very common in the hills around the Italian city of Verona and it contains many ammonites, which can be spotted walking around the city, in fountains or monuments, in the floors of the city's palaces, or in the breakwaters of the many villages of nearby Garda Lake. Occasionally, sport climbing crags near abandoned quarries in the area allow the rock climber a sensory experience of these ammonites. More in general, *ammonite red* is the name given to the deposits of sedimentary rocks formed in the Jurassic period around 185 million years ago, which contain the spiral-shaped shells of large molluscs very similar to the species of Nautilus still existing today. The name comes from these ammonite fossils and from the markedly red colour.

In addition to fossilized plants and animals, there are petrified wood and watermarks made on sand which later fossilized. Footprints in mud or sand can also become fossilized and a few "dinosaur trails" in the mountains allow one to see dinosaur footprints in the days off climbing because of sore limbs or iffy weather. Marine fossils and shells are often found on the top of sedimentary mountains, a testimony of the powerful forces which raised layers of sediments deposited at the bottom of a tropical sea to the now cold slopes and summits. Fossils allow the amateur to do science beyond the amateur level. If you are in Cortina d'Ampezzo, in the heart of the

Italian Dolomites, don't miss the Zardini paleontological museum. Rinaldo Zardini (1902–1988) was a local photographer, naturalist, and botanist. Elevating himself from the ranks of self-educated collectors, he was affiliated with the Smithsonian Institution in Washington as a researcher and was responsible for assembling collections of fossils from the Dolomites, many of which he found himself. He is said to have collected one million specimens. Today the Zardini museum hosts one of the most complete collections of fossils from the Triassic period.

Metamorphic rocks originate from igneous or sedimentary rocks. They are rocks which are buried deep into the Earth and are subject to intense heat and pressure, which reprocess them. Igneous or sedimentary rocks subject to these agents are changed, also in their crystalline structure (they re-crystallize), and become metamorphic rocks. Metamorphic rocks include slate, phyllite (compressed and heated slate), schist (compressed and heated phyllite), gneiss (compressed and heated schist), marble (squeezed/heated limestone), serpentinite and amphibolite.

Rocks don't last forever. On geological time scales, rocks are processed in a rock cycle which includes burying them deep inside the Earth, where they are subject to heating and to enormous pressures. They can melt partially or totally and are reprocessed to finally form different types of rock. They may disappear under a tectonic plate and melt, or be pushed toward the surface, or emerge through a volcano, or being slowly exposed by erosion, sometimes making the enjoyment of rock climbers.

1.7 Destroying mountains

Erosion is a process of destruction of the top parts of the lithosphere and some forms of erosion are particularly evident in alpine terrain. Erosion can be due to mechanical, chemical or (not important in the mountains), biological action (Fig. 1.6).

Mechanical factors, which usually prevail, include the action of glaciers responsible for moraines and U-shaped valleys, of running water creating V-shaped valleys, deep gorges (Figs. 1.1 and 1.2), and

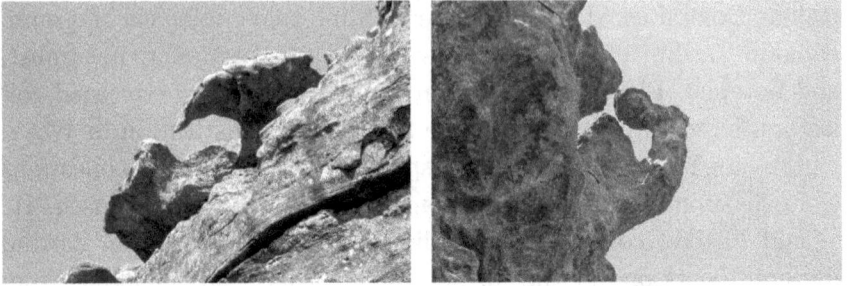

Fig. 1.6　Quartzitic sandstone sculpted by erosion (Table Mountain, South Africa).

channels on scree slopes, glaciers, and rock, plus frost weathering and eolic action.

Glaciers which carve mountain sides often give them pyramidal shape and create U-shaped valleys. Very famous in mountain culture is the shape of the Matterhorn on the Swiss–Italian border which made climbing it very hard in the early history of mountaineering [Newby (1977)] and attracts mountaineers from all over. Indeed, this mountain defines the "mattherhorn shape" and various mountains with this shape are known around the world. Mt. Assiniboine in the Canadian Rockies is another example. Glaciers move materials, grinding the rocks below and around them and pushing gravel, rocks, till, boulders, and other materials, while forming frontal and lateral moraines.

Erratic boulders are isolated rocks that do not belong to the local geology and look as if they were deposited there by some giant hand. It was the hand of a glacier that carried them from far away and left them behind when the ice melted. Erratic boulders are greatly appreciated by rock climbers for the bouldering they provide in flat plains with no rock walls around.

Frost weathering is due to the action of water which penetrates deeply into cracks and crevices, and then freezes and expands. The stresses applied to the surrounding walls are large and fracture the rock. When it melts, liquid water penetrates deeper into the crack, then freezes again, and the cycle repeats. Large boulders can be fractured this way (Fig. 1.7).

Fig. 1.7   Frost weathering of granite on a mountain top.

Running water erodes slopes and collects to form small streams that progressively grow larger and larger. The steeper the slope, the faster the water runs[11] and the more pronounced the erosive action (given equal material hardness). Moving water carries materials which deposit as the slope lessens, milder angles are approached, and the speed of the water decreases. It is said that running water transports materials and sorts them out by particle size. Larger materials (boulders and bigger rocks) may travel only a little and are deposited first, followed by pebbles, sand, silt, and clay which may continue their journey downstream to flat alluvial plains. Large rivers, such as the Amazon or the Mississippi transport fine-grained materials all the way to the oceans.

Rivers create V-shaped valleys. Also running water in the mountains transports several materials, ranging from small boulders to

---

[11]Neglecting friction, at constant acceleration $g \sin \theta$ (where g is the acceleration of gravity and θ is the slope angle) a particle (or a falling climber who doesn't self-arrest with an ice axe) picks up a speed $v(t) = gt \sin \theta$ as the time t goes by.

pebbles and finer sand and clays in suspension. Water can cause slides, especially when it saturates the soil. On glaciers, running water can excavate deep runnels which, full of frigid water and with rounded and slippery walls, are dangerous places for mountain climbers and end in crevasses or moulins.

Since rocks are poor heat conductors, rapid thermal variations occur only near their surface. Thermal dilation and contraction occur mostly near the surface and, depending on the type of rock, can cause exfoliation over time. Rocks absorb heat differently according to their colors and can dilate differently in different directions, due to anisotropic crystalline structure, when a crystal dilates at different rates along different axes.

Wind erosion is dominant only in deserts and dry lands, where other forms of erosion are absent or largely irrelevant. Wind erosion can create rock sculptures familiar to rock climbers who winter in the deserts of the American Southwest or to tourists vacationing on the Sardinia island of Italy, in Corsica, or in other islands exposed to sea winds where wind erosion is significant. Eolic erosion is due to the grinding action of hard particles carried by the wind. When snow is present, the most relevant action of the wind is to transport snow which, deposited on the lee side of slopes, must be noted in order to properly assess avalanche hazard (Sec. 3.9). In sandy deserts, wind can transport entire sand dunes (which are also known to be fun for occasional skiers).

Chemical erosion, which takes place mostly in humid areas, is due to the presence of water, oxygen, and carbon dioxide and is markedly different according to the type of rock eroded. Art conservationists know well the significant difference between the weathering of ancient sculptures made of granite (for example, those found in the ancient Indian city of Hampi) and those made of limestone (for example in Angkor Wat, Cambodia).

Karstic rock[12] owes its distinctive features to chemical erosion. Water (H_2O) and carbon dioxide (CO_2) combine to form the weak

---

[12]Karstic rock derives its name from the Karst region shared by Slovenia and Italy near the city of Trieste, a heaven for cavers.

carbonic acid (H_2CO_3) in the chemical reaction

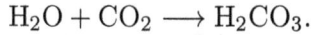

$$H_2O + CO_2 \longrightarrow H_2CO_3.$$

The carbonic acid produced then reacts with limestone ($CaCO_3$) and dissolves it in the other chemical reaction

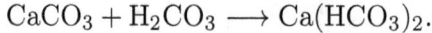

$$CaCO_3 + H_2CO_3 \longrightarrow Ca(HCO_3)_2.$$

The calcium carbonate $Ca(HCO_3)_2$ dissolved in water is deposited when water drips slowly and meets air in caves, forming stalactites and stalagmites over the course of centuries (Fig. 1.8).

The chemical reaction above is responsible for the karstic land-scape characterized by caves, sinkholes, and all kinds of holes and channels separated by sharp ridges (*rillenkarren* or solution flutes) on the surface of the rock (Figs. 1.9 and 1.10). They make wonderful climbing holds when they appear in the right sequence and position and sometimes even host natural bridges for protecting the climb. If rillenkarren become too large, however, their sharp ridges constitute a threat to a climbing rope.

Holes due to dissolving limestone can be interesting. Sometimes, behind a hole in limestone there is a cavity carved by water which is larger than the hole and may have complicated geometry, perhaps connected to other holes piercing the surface of the rock through an intricated system of small tunnels which do not attract only climbers. I know a rock climbing route on a limestone cliff which has a vertical sequence of holes, and in each one of them used to reside a dormouse. The entire rodent family would get agitated when a climber would arrive to that point of the route and would begin sticking fingers in their houses. On another favourite sport climbing route, the only handhold available on an otherwise per-fectly smooth eighty-degrees limestone slab was a deep two-finger pocket which, at times, hosted a rodent. Getting two fingers into the pocket required a very reachy and committing move, which was fol-lowed by an unfriendly growl coming from inside the hole. Although I never heard reports of a bite, it was still unpleasant and very weird. But still better than the experience of that fellow climbing in the Karst. He started climbing a route which begins with a small

Fig. 1.8   Stalactites and stalagmites (Postojna, Slovenia).

cave, past which one uses a triple hole to gain height and proceed. When he arrived there, a viper came out of one of the holes. At least, the vipers that I met climbing limestone routes on two separate occasions were out in the open and not hiding in deep holes.

Fig. 1.9   Rillenkarren on limestone (Monte Baldo, Italy).

Fig. 1.10   A hole and small rillenkarren caused by erosion in soft limestone.

Who knows what's inside and behind all those pockets created by carbonic acid!

Biological erosion in the mountains is limited to the action of lichens, moss, roots, bacteria, and algae and is usually negligible in comparison with the other, much more pronounced, forms of erosion.

1.8   Of sweat, rock, and hard places

Traditionally, the minerals composing rocks are classified according to their chemical composition, but mountaineers and climbers may be more interested in their hardness, which determines the landscape, affects the likelihood of loose rocks falling on them or of holds breaking off, and the solidity of anchors placed in the rock. Very few people like to climb on choss and runouts on loose alpine rock are downright scary. From the physical point of view, minerals are also classified on a scale of hardness based on the empirical criterion that a mineral is higher on this scale if it scratches another material of known hardness (lower on the scale). Hardness scales for materials were introduced in antiquity by Theophrastus, the Greek philosopher successor to Aristotle, and by Pliny the Elder, the great Roman investigator of nature and geography, who apparently climbed some mountains, too, long before the sport of mountaineering was invented, and who died in the eruption of Mt. Vesuvius in AD 79. The hardness scale used nowadays was introduced by the German scientist Friedrich Mohs (1773–1839) and is now called the Mohs' scale. Mohs, the most famous mineralogist of his times, was also a collector of minerals and died in Agordo, in the Italian Dolomites, during a collecting trip.

The Mohs scale goes from H1 or hardness 1, corresponding to talc (the softest mineral) to H10, corresponding to diamond (the hardest mineral), as shown in Table 1.1.

The scale admits half-grades. The hardness of the minerals prevailing in rock formations determines certain features of the landscape. For example erosion, the great destroyer of mountains, creates large piles of scree at the bottom of the walls and towers of the Dolomites (Figs. 2.3 and 2.4), more so than in harder igneous rock, because dolomite rock has a Mohs hardness 3.5–4, relatively low on

Table 1.1   The hardness of various
materials on the Mohs scale.

Material	Hardness
Talc	H1
Gypsum	H2
Calcite	H3
Fluorspar	H4
Apatite	H5
Orthoclase	H6
Quartz	H7
Topaz	H8
Corundum	H9
Diamond	H10

the scale. Limestone mountains also erode easily and are associated with huge piles of scree. Limestone has hardness H3–H4 and marble has H3–H5. For comparison, granite has hardness H6–H7. A note to this regard for sport climbers: while (leaving aside ethics and the violence to lichens) it seems acceptable to wire-brush a granite sport climb to clean it, one should never do that on a limestone climb or one risks ruining the holds and facing angry local climbers. Limestone is much softer than granite and a plastic brush should be used on limestone instead of a metal wire brush.

On a grander perspective, one notes that planet Earth is about 4.5 billion years old. The oldest known exposed rocks, made of Acasta gneiss, were found in the Northwest Territories of Canada and are almost four billion years old. Gneiss is formed deep within the Earth's crust and is seen on mountains where the rock has been uplifted, or where erosion has removed their cover (for example, on the Canadian Shield). Likewise, old rocks can be found at the bottom of deep gorges like the Grand Canyon, which is not surprising because, for a geologist, a trip down the Grand Canyon is equivalent to a time travel to the past.

Limestone is a sedimentary rock that makes limestone mountain ranges and cliffs considered to be the heaven of sport climbers. It often incorporates the skeletons of ancient marine organisms and is usually composed of calcite and aragonite. These minerals are two crystalline forms of calcium carbonate ($CaCO_3$). As already discussed in Sec. 1.7, limestone is easily dissolved by carbonic acid and other weak acids present in rain and forms characteristic "karstic" landscapes full of holes, caves, and depressions caused by the dissolution of the rock. A note for sport climbers who are limestone aficionados: sweat is acidic and the climbing chalk used to dry one's hands, which is basic, partially neutralizes the acid and slows down corrosion. This is not a small issue in popular limestone climbing areas since routes there quickly become polished. There are, however, claims that chalk accelerates the deterioration of very porous sandstone which absorbs it. Because of limestone's propensity to dissolving, it is good practice to be especially careful and to check frequently bolts on limestone cliffs which receive a lot of rain, such as those in tropical areas.

1.9   Volcanoes

Climbing volcanoes can be doubly satisfactory because of the panorama around the volcano and of the view inside (if you can get them). In addition to the usual thrill of climbing a mountain, one gets unforgettable views of the primeval forces of nature at work and of the direct products of underground hellish fire (sometimes, of the fire itself). Many volcanoes are high and heavily glaciated mountains, which makes them even more attractive for mountaineers.

Magma making it to the surface of the Earth through volcanoes can be granitic or basaltic. Granitic magma comes from depths less than about 15 km and is cooler than deeper magma, contains more than 65% of SiO_2, is acidic and quite viscous. Because of these properties it tends to flow less and to solidify earlier than basaltic magma, and it tends to form plugs in the chimneys of volcanoes. These plugs allow the accumulation of gases underneath and the buildup of high pressure, until these gases are released suddenly like the cork of a champagne bottle when the plug is removed. Granitic magmas then give rise to explosive volcanoes, or *stratovolcanoes*, in which eruptions are marked by explosions and by the release of large amounts of ash and gas and by the projection of volcanic bombs. Explosive volcanoes on land[13] are characterized by high and prominent cones. Examples of stratovolcanoes are Mt. St. Helens in the Northwestern United States, Mt. Etna in Sicily, Mt. Vesuvius near the city of Naples in Italy, Krakatau in Indonesia, and Mt. Fujii in Japan. Some of these names are associated with catastrophic historical eruptions causing, among others, the obliteration of the Roman cities of Pompeii and Herculaneum by Mt. Vesuvius in AD 79, the tsunamis unleashed by Krakatau in 1883, and the brief climate change due to the powerful eruption of Mt. Tambora in 1815 and to the prodigious amount of volcanic ashes erupted.

By contrast, basaltic magma originates at larger depths between 15 and 75 km, is hotter, basic, contains less than 52% of SiO_2 and little in terms of dissolved gases, has low viscosity, and flows well. It gives rise to *shield volcanoes* shaped like the Hawaiian volcanoes in which lava flows out in lava fountains without explosions, and cones are low and with mild slopes. Being hotter, basaltic magma takes longer to solidify than granitic magma and runs longer along slopes.

When a lava flow encounters a body of water (lake or sea), it solidifies abruptly, contracting and taking columnar shape, which a long time later can make for good trad climbing.

Volcanoes are found in many places around the world but are especially evident along the ring of fire of the Pacific Ocean and along oceanic ridges. The west coast of North America is dotted with volcanoes. Starting in Alaska and in the Aleutian Islands (the site of fifty eruptions since the mid 1700s) and descending South along British Columbia, one encounters Mt. Edziza, the Wells Gray volcanic field, Mt. Silverthrone, the Nazko Cone, Mt. Garibaldi and the Black Tusk, and Hoodoo Mountain, some of which are popular with mountaineers from the Northwest. The series continues through the Cascades ranges in Washington State, Oregon, and California where one encounters Mt. Baker, Glacier Peak, Mt. St Helens, Mt. Rainier, Mt. Hood, Mt. Adams, Mt. Shasta, and Lassen Peak. The series

---

[13]There are also submarine eruptions, which do not seem very relevant for mountaineers.

of popular climbs continues with the volcanoes around Mexico City (in particular Paricutin, Nevado de Colima, Popocateptl, Pico de Orizaba, and Ixtaccithuatl) and proceeds through Central America, notorious for seismic and volcanic activity, to continue through South America where Chimborazo, Huascaran, and Cotopaxi are lofty mountaineering goals. Across the Pacific along oceanic ridges, the ring of fire continues to Russia, Japan and South to the Philippines, Indonesia, Java (Krakatau, Mt. Tambora, Mt. Pinatubo, Mt. Mayon, Kelud, Merapi are well known), New Zealand and the South Pacific to the Tonga and the Kermadec trenches, and down to Antarctica.

Chapter 2

Gravity rules

2.1 Introduction

An old T-shirt from a manufacturer of climbing equipment stated that "the law of gravity is strictly enforced". Indeed this fact is nowhere more evident than in alpine terrain. Gravity rules in the mountains. Downward mass movements induced by gravity (*mass wasting*) can be very remarkable and include avalanches, landslides, debris flows, rock slides, boulders tumbling down a hill, and rocks falling off vertical faces. The pull of gravity seems stronger to a climber on a vertical wall, and enormous on overhangs. Climbing up a steep slope of ice or hard snow, one is always aware that a slip means gaining speed very quickly because of that gravity. And gravity is also felt just hiking up a steep slope, especially if covered in fine scree that gives way and results in one step down for each two steps up. The same pleasure is experienced by postholing up to one's waist in soft snow on steep slopes, which brings us to inclines, a classic subject of first year physics courses. One of the goals of these physics lectures is to make students realize that a force is a vector, a quantity which has a direction associated with it, in addition to a magnitude. Gravity always points vertically and, obviously, down, but when we are on that steep North face snow slope this means that our weight (a vertical arrow) decomposes into two directions.

The first component is perpendicular to the slope and is balanced by the slope itself, according to Newton's law of action and reaction (after we stop sinking in the soft snow, that is). The other component of the weight force points parallel to the slope, and that's the one we worry about on a hard, slippery surface. On hard snow or ice one doesn't sink and all one cares about is this component, which persistently wants to take us for a ride down the slope. That's why carrying crampons in snowy mountains is never pointless.

The component of the weight pointing along the slope is not always a bad thing: it is the essence of skiing. Bootskiing down a firm snow slope is also fun and a fast way to get down a summit quickly for a climber who doesn't carry skis and doesn't trip. And if one trips, self-arrest is the first ice axe skill learned, which avoids picking up a dangerous speed on a snow slope. Without friction, the acceleration along the slope is $g \sin \theta$, where θ is the slope angle and $g = 9.8 \, \mathrm{m/s^2}$ is the (vertical) acceleration of gravity (Fig. 2.5). When crossing a rocky slope, an experienced climber tends to maintain a vertical position above the holds. Leaning into the slope, as many beginners do, gives a false sense of security and makes it easier to slide (Sec. 6.2). It is much better to keep vertical above the footholds and to minimize this component. Experienced climbers feel naturally this decomposition of the vector force acting on their bodies.

Gravity pulls down rocks that accumulate to form talus cones at the base of walls; it pulls down water which runs in streams in summer and carries debris of various sizes. In fact, nothing obeys the pull of gravity as quickly as water, which doesn't have edges that catch on obstacles and rubs only a little against surfaces (enough to carve V-shaped valleys, though). Water takes the steeper path, which is why it is generally a bad idea to follow a stream when descending in unknown terrain. Some of my West Coast friends seem to take pride in ignoring this rule and point out that this method often works to take you down quickly. I admit that it has worked for us on a couple of trips but it was not fun other times when finally, and predictably, we got stuck and had to backtrack uphill. In other situations, such as when canyoning, the game consists of following the water along the steepest path, as in Fig. 1.2.

Fig. 2.1 Avalanche paths (Banff National Park, Canadian Rockies).

Gravity causes avalanches by pulling large quantities of snow along slopes and gullies: avalanche runs are evident in most mountains (*e.g.*, Fig. 2.1). Gravity causes also rock and mud slides, which are more frequent in the mountains because of the steep terrain. Seracs fall relentlessly from icefalls, and cornices fall from ridges when their internal stresses fail to hold them up. In short, gravity defines the steep mountain environment, is the source of most of the climber's challenges, but also guarantees the psychological rewards for overcoming these challenges. Let's look at a few issues associated with the eternal struggle of the mountaineer against gravity.

2.2 How hard will a falling rock hit?

Falling rocks kill, even small ones if they fall from higher up. When climbing a long multi-pitch rock route it is scary to hear the hissing sound of a rock falling fast through the air, coming from two or three hundred meters higher and this can happen several times during a climb. How hard will a falling rock hit? In physical terms, this question is better translated into "how fast can a falling rock move?" How much energy will it acquire and how much force will it apply when it hits?

Let us begin with a rock of mass m falling from a height h measured from the site of the impact, hopefully the base of the wall far away from us. Before beginning its fall, the rock has energy, called gravitational potential energy, simply due to the fact of being up high. The amount of this energy is mgh, where m is the mass of the rock and $g = 9.8 \, \mathrm{m/s^2}$ is the acceleration of gravity (we assume that the rock does not have initial velocity). If we neglect air friction, which is only justified for short falls, after falling a height h the rock will acquire a speed v and a kinetic energy (energy of motion) $mv^2/2$ equal to the initial gravitational potential energy:

$$mgh = \frac{mv^2}{2}.$$

This is because, neglecting air friction, energy is conserved and the total energy (potential plus kinetic) is constant. As the rock hurls down, gravitational potential energy is converted into kinetic energy. In other words, whatever potential energy the rock loses by coming down and reaching lower elevations, it gains as energy of motion by going faster.

You see that the mass m of the rock cancels out in the previous equation because it appears in both sides with the same power, so it disappears from our calculation from now on (this is assuming that only gravity, but not air friction, is relevant, which is what led us to this equation). Then, we obtain the velocity of the rock

$$v = \sqrt{2gh}.$$

To get an idea of what kind of velocity can be reached, think of a rock falling for 100 meters, which gives

$$v = \sqrt{2\left(9.8\,\frac{\mathrm{m}}{\mathrm{s^2}}\right) \cdot (100\,\mathrm{m})} = 44\,\mathrm{m/s} = 160\,\mathrm{km/h},$$

where we used the conversion factor[1] $1\,\text{m/s} = 3.6\,\text{km/h}$. It is a good idea to get out of the way of any falling rock and to always wear a helmet on alpine climbs.

In actual fact, air friction cannot be neglected for long since it decelerates the rock and, if the fall lasts long enough, it may even balance the force of gravity. The air drag depends on the speed v: at small speeds one can assume that this air drag is proportional to the speed v, while at large speed a physicist will tell you that it goes like v^2. Let us assume that this force is $F_{\text{drag}} = Dv^2$, where D is a constant, which depends on the shape of the rock, and is called *drag coefficient* in slang. If the rock keeps falling, gravity will accelerate it only up to a maximum speed v_*, at which the drag force balances the force of gravity. At this point the rock is subject to zero net force and, from now on, it has zero acceleration, that is, from that point on it will fall with a constant speed called *terminal speed*. This terminal speed v_* is computed by balancing the two forces:

$$Dv_*^2 = mg,$$

which gives

$$v_* = \sqrt{\frac{mg}{D}}. \tag{2.1}$$

Contrary to the velocity acquired in free fall, the terminal velocity in the presence of friction does depend on the mass of the falling rock. If the mass of the rock is, say, $m = 1.0$ kg (a relatively small rock) and $D \simeq 0.5\,\text{kg/m}$, one obtains a terminal velocity $v = 4.4\,\text{m/s} = 16\,\text{km/h}$. Equation (2.1) tells us that, if the mass m is much smaller (for example, for water droplets in fog), or if the drag coefficient D is much larger (for example for a paraglider,[2] as in Fig. 2.2), the terminal speed could be much less.

---

[1] $1\,\text{km/h} = 1000\,\text{m}/[60\,\text{minutes} \cdot (60\text{s/minute})] = (1/3.6)\,\text{m/s}$ and, therefore, $1\,\text{m/s} = 3.6\,\text{km/h}$.

[2] Paragliders do not just fall, but they take advantage of ascending air currents caused by temperature gradients to gain elevation again.

Fig. 2.2   The drag coefficient is very large and the terminal speed is very low for this paraglider (La Tournette, France).

The next question is: how much force will a rock falling with velocity v apply when it hits something? We need some more mechanics. The momentum of an object is $p = mv$ and Newton's second law of motion relates the applied force F with the variation of momentum Δp and the length of time Δt over which it is applied,

$$F = \frac{dp}{dt} \simeq \frac{\Delta p}{\Delta t} = -\frac{mv}{\Delta t},$$

assuming that the rock comes to rest upon impact, in which case the variation of momentum of the rock is

$$\Delta p = p_{\text{final}} - p_{\text{initial}} = -p_{\text{initial}} = -mv.$$

The duration of the impact is quite short, say $\Delta t = 0.005$ s which gives, for our 1.0 kg rock at terminal speed,

$$|F| = \frac{mv}{\Delta t} = \frac{(1.0\,\text{kg}) \cdot (4.4\,\text{m/s})}{0.005\,\text{s}} = 880\,\text{N}.$$

This is equal to the weight of a $880\,\mathrm{N}/g \simeq 90$ kg mass. This force is applied on a small contact area A, generating a large pressure. If the rock has sharp edges, which is likely for a rock fractured by freezing and thawing on a mountain, this contact area can be very small. Say, for illustration, that this is $1\,\mathrm{cm}^2$, then the pressure generated is

$$P = \frac{F}{A} = \frac{880\,\mathrm{N}}{10^{-4}\,\mathrm{m}^2} = 8.8 \cdot 10^6\,\mathrm{Pa},$$

the equivalent of $\sim$88 atmospheres ($1\,\mathrm{atm} = 1.01325 \cdot 10^5\,\mathrm{Pa}$). Since the impact of rocks with different types of materials of varying strength requires many details to be described, these figures are purely for illustration and can vary greatly, but they nevertheless give an idea. One should always wear a helmet below or on a rock or ice wall. When climbing in a group, apart from the obvious suggestions of taking care not to dislodge rocks and not staying one below the other if possible, it pays to stay fairly close together to minimize the height h that a rock falls from, its potential energy mgh, and the speed $\sqrt{2gh}$ that it acquires. If rocks fall from a very high place, it is common sense to choose the climbing route and timing the climb in such a way to eliminate or minimize the danger of rock or ice falling and to limit the exposure to these objective hazards as much as possible. This often includes climbing a route at night (cf. Sec. 3.6).

2.3 Talus cones, avalanches, and antlions: the angle of repose

Remember those steep talus cones on which the trail was climbing to get to the last summit? Hard to go up on the fine scree, one step up and one step down as the unstable slope gave way. It seemed it would never end. Way easier on the way down, when the same fine scree made for an exhilarating run without effort. Two hours to go up and barely twenty minutes to come down. A similar story can be told about that final slope of super-dry volcanic ash on Glacier Peak in the North Cascades: it wasn't scree and it wasn't as steep, but it was still unstable and a long, long way up (which seemed unfair once one was above the glaciers), but a fast way down. How steep can those scree or ash slopes get? Certainly the interlocking pieces of rock can create steeper slopes than ash (Figs. 2.3 and 2.4).

Physicists have something to say about these slopes and they express it in terms of the *angle of repose*. This term has two distinct, although not unrelated, meanings in physics. The first one is familiar from elementary mechanics courses. Suppose that you have an inclined plane (an *incline*) that you can tilt at will to adjust its slope angle. Put a block of some material on the incline when it is horizontal and then tilt this plane slowly. Because of friction (in this case *static friction* because the block doesn't move), the block will stay in place. Continue increasing the slope until the block starts sliding down. The critical angle when the block releases and just begins to slide down the slope is called *friction angle* or *angle of repose*. Let θ be the angle that the incline makes with the horizontal, then the weight force of the block (a vector, represented by an arrow) points

Fig. 2.3   Talus cones near the angle of repose.

Fig. 2.4   Scree slopes in the Italian Dolomites.

vertically and has intensity mg, where m is the block mass and g is the acceleration of gravity (Fig. 2.5).

We decompose this vertical force into two component forces (or *components*), one pointing perpendicular to the slope and one parallel to it. According to Newton's law of action and reaction

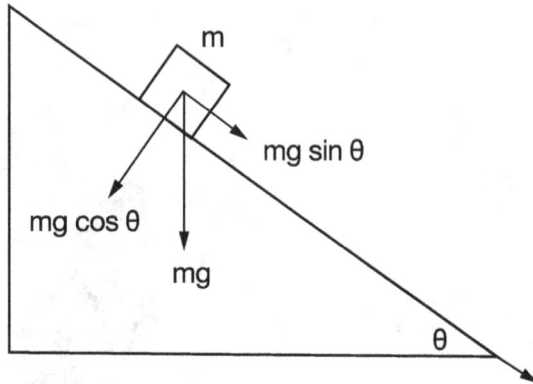

Fig. 2.5   A block on an incline — decomposition of the weight in components
parallel and perpendicular to the incline.

(when a body A applies a force $\vec{F}$ to another body B there is an
equal and opposite reaction $-\vec{F}$ applied by B on A), the first component is balanced by the reaction of the plane, so we only need to
deal with the component of the weight parallel to the slope. That's
the one which pulls us down, wonderful if we are skiing, less so if we
are climbing a steep North face slope on ice. Good old trigonometry
tells us that the components parallel and perpendicular to the slope
have intensities $F_{\parallel} = mg\sin\theta$ and $F_{\perp} = mg\cos\theta$, respectively. This
is just geometry: now physics tells us that the friction force, which
balances the weight component along the slope until this becomes
too steep, has intensity

$$F_{\text{friction}} = \mu_s F_{\perp} = \mu_s mg\cos\theta,$$

where μ_s is a pure number (that is, without dimensions or units), the
coefficient of static friction. When the block remains in equilibrium
on the slope, friction balancing the component of the weight parallel
to the slope, it is

$$F_{\text{friction}} = F_{\parallel}$$

or

$$\mu_s mg\cos\theta = mg\sin\theta.$$

Solving for the angle θ gives

$$\tan \theta \equiv \frac{\sin \theta}{\cos \theta} = \mu_s. \tag{2.2}$$

θ is the friction angle. This equation is used to measure and tabulate the friction coefficients for various materials in contact by changing the composition of the block or the incline.

When the block starts moving (it could be a skier on a slope instead), friction is no longer static but it becomes dynamic. The friction force always opposes the motion but in this case it is not sufficiently large to balance the component of the weight $F_{\parallel} = mg \sin \theta$ pointing downhill. If you picture a coordinate axis pointing downhill along the incline (Fig. 2.5), gravity causes forward motion and $F_{\parallel}$ is positive, while friction opposes the motion and is negative, $F_{\text{friction}} = -\mu_k mg \cos \theta$. Here μ_k is the coefficient of kinetic friction, a pure number distinct from μ_s which quantifies friction in dynamical situations. Newton's second law of motion says that (total) force equals mass m times acceleration a and we use it here, obtaining

$$ma = mg \sin \theta - \mu_k mg \cos \theta$$

in the direction parallel to the incline. The masses cancel out, a property unique to the gravitational force, which applies not only to free fall but also to inclines in which only one component of the weight causes the acceleration.[3] The acceleration of the block (or skier) along the incline is then

$$a = g \left(\sin \theta - \mu_k \cos \theta \right),$$

a fraction $(\sin \theta - \mu_k \cos \theta) < 1$ of the acceleration g that the object would have in free fall.

The second meaning of angle of repose is more sophisticated and refers to piles of granular materials including sand, snow, gravel and,

---

[3]It may or may not be a myth that Galileo Galilei first studied gravity and discovered the universality of free fall by dropping objects from the Leaning Tower of Pisa, as reported by his biographer Viviani, which has been the subject of a long debate [Segre (1989)]. Almost certainly, however, he used inclines in his experiments.

by extension, piles of scree and talus cones. The angle of repose is the maximum inclination that a part of the talus can maintain without collapsing (the largest slope angle). The angle of repose refers to granular material on the verge of collapsing. It is intuitive that, in nature, a slope close to the angle of repose won't last for very long because there will be some perturbation (avalanche, falling rocks, snow load, wind, an animal crossing the slope, *etc.*) that will perturb it and make it collapse. Avalanche hazard sets in when snow accumulates near the angle of repose and avalanches are triggered when this angle is exceeded, either because of extra load or of weakening bonds in the snowpack.

I remember finding a scree slope very near its repose angle once. The peaks of Vancouver Island in Western Canada have been shattered by earthquakes and some of them have piled up an unusual amount of debris in a, geologically speaking, very short time. I was reaching the summit of Elkhorn Mountain when, in spite of being very careful of how I moved because of the loose rock, I knocked down a pebble the size of two or three centimeters from a step a couple of meters high. Elkhorn Mountain is known as "the Vancouver Island Matterhorn" because of its imposing profile from far away but, aside from its modest elevation of 2194 m, is nothing like the much more compact Mattherhorn. The summit of Elkhorn is a steep pile of loose rock. A historical debate about it being the highest peak of the island has been settled long ago with the conclusion that Elkhorn is about one meter lower than the island's highest peak, the Golden Hinde. However, Elkhorn might have been the highest peak before its summit was shattered and reduced to a pile of loose rock by an earthquake [Elms (1996)]. Which brings us back to my pebble: it bounced off moving a couple of little rocks, then bounced again and moved a little further down, and then again, and so did the two little rocks, and the entire slope suddenly moved down in a cascade process. The entire slope was clearly at a critical point. I have never seen anything like that before, so unstable: it was sitting at the angle of repose and just waiting to go. Fortunately, by now my climbing friends were above the slope and contouring a band of more solid rock and there were no consequences, but I felt bad for months afterwards.

Normally scree slopes adjust spontaneously and do not wait for our help to do so. Snow slopes sometimes do and the angle of repose is used to assess avalanche hazard. Only, mountain slopes are not perfect smooth planes and snow comes in a nearly infinite variety of forms with different grains and bond strengths which can't be captured in a single simple formula like Eq. (2.2) for the angle of repose of a block on an incline. In practice repose angles can vary, something around 25–30 degrees for gravel, 35–40 degrees for snow, 45 degrees for chunks of granite.

Interestingly, antlions know everything about the angle of repose. They dig conical pits in loose sand, with the slopes of the funnel built at the angle of repose. When an ant or other insect wanders on one of these slopes, the latter collapses bringing down the unlucky creature. The process is helped by the knowledgeable antlion which hides at the bottom of the pit and flicks sand grains on the walls to make sure the process works well. Goodbye ant and thanks to both for the lesson.

2.4 Water erosion

Gravity does a lot to destroy mountains and water greatly helps. Like anything else, water is pulled by gravity and, because it doesn't have a shape or edges catching on obstacles, it picks the steepest path to arrive to lower elevations. The momentum of a rushing stream can move big boulders and, likewise, the momentum of water in smaller streams moves smaller rocks, debris, and other materials which in turn erode the bed and the walls of the channels in which they travel. These materials scrape far more than the water does. The steeper the slope, the faster the motion, the higher the kinetic energy of the water and of the materials moving with it, and the more mechanical energy is dissipated in friction against the bed and the walls of the channel.

V-shaped valleys and deep gorges are testimonies of monumental proportions to the erosive power of water. Erosion happens on sheer rock, and much more so in soft soil. When soil is present, its weight is increased by the presence of water, but running water also erodes easily soft terrain and decreases friction by lubricating the surface of contact between soil and bedrock and between soil particles themselves, which can result in reduced cohesion and massive landslides after heavy rainfalls. Sometimes the physics goes in the opposite direction: a pile of dry sand is less cohesive than wet sand and has a smaller angle of repose, but if the sand is completely permeated by water it becomes more unstable and flows almost like a liquid (if it is on a horizontal surface and it is covered by a layer of more solid soil, it may become quicksand [Boeker and van Grondelle (2011)]). When present, vegetation acts to stabilize a slope with its roots. The deforestation of mountain slopes makes them particularly prone to landslides, especially if the soil is thin. This circumstance is particularly evident in the rugged mountains of coastal British Columbia where temperate rainforest has been clearcut and rainfall is abundant. The effects of indiscriminate mountain logging were obvious in Honduras and Nicaragua when Hurricane Mitch struck in 1998 with extremely heavy rainfall, causing huge mudslides, the death of thousands, and massive destruction.

The freeze-thaw action of water penetrating into cracks is also a major agent of erosion in the mountains, especially during spring and fall and at dawn and dusk, when the temperatures oscillate around zero degrees Celsius (Sec. 3.6 and Fig. 1.7).

On a much smaller scale, water erosion is always present where it rains. Park crews in charge of trail maintenance know this well: a badly designed trail in steep mountain terrain with significant rainfall does not last for long. It is especially important to provide channels for the water to run through across the trail, otherwise the trail itself will become a stream and rainwater will dig a channel which soon makes sections of the trail difficult or impossible to use. Water falling on the surface of steeper snow only partially penetrates it and creates runnels which are quite obvious even after the rainfall (Fig. 2.6).

On the surface of glaciers, meltwater runs in channels which end in crevasses or in *moulins*. Moulins or glacier mills are holes of approximately circular shape which are vertical or nearly vertical and carry

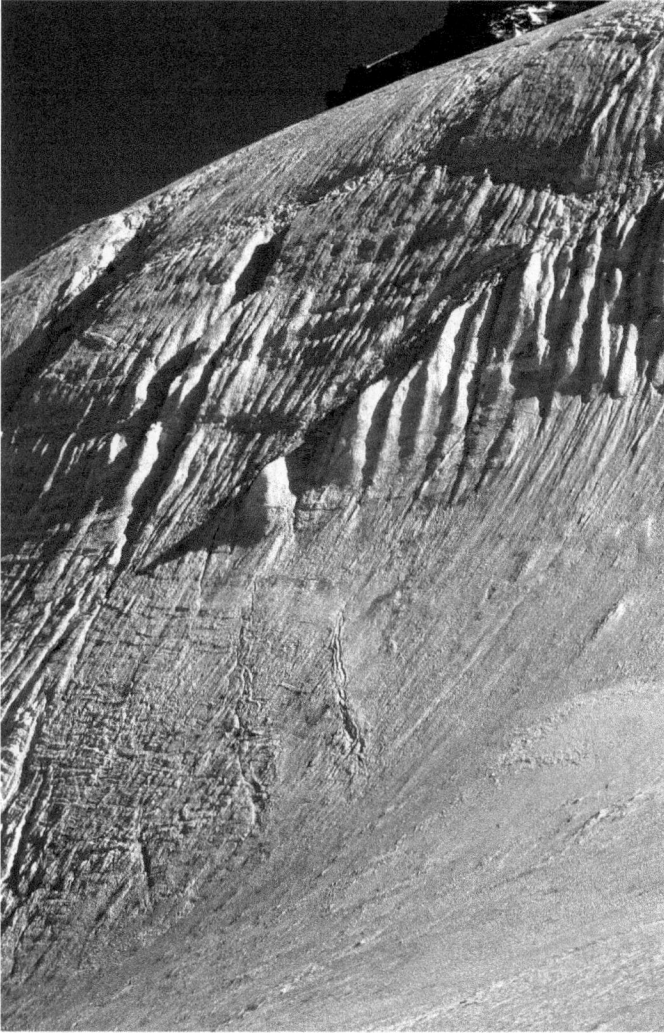

Fig. 2.6   Runnels carved by water in steep snow.

meltwater deep into the glacier, similar to shafts in a mine. They are part of the complex hydrology of glaciers.

When a mountain stream or river flows to progressively gentler terrain, materials are deposited according to their size, as the horizontal force of the water becomes less and less in comparison with the weight. Boulders are deposited first, then large rocks, then pebbles, and finally sand. When the mountain river makes it out to the plain, finer and finer materials are still present. Silt and clay may be deposited a long distance from the sources, while new materials may be injected in the flow by erosion caused by the river. Large rivers such as, for example, the Amazon or the Mississippi carry these finest materials all the way to the ocean.

2.5 Sequoias and capillarity

Water rises by capillarity through the walls of buildings and through natural and artificial fibers, but how high can it really rise? Often a mountain climb begins below treeline (in some regions of the North American Pacific Northwest it may never leave it) and one may walk for hours through high timber. In California and Oregon one may cross sequoia woods with ancient trees reaching as high as 80 m. Can capillarity rise water to the leaves which are at the top of these giant trees? To answer this question one needs to know that the height above the free surface of a liquid which is reached in a capillary tube is [de Gennes, Brochard-Wyart, and Quéré (2010)]

$$h = \frac{2\gamma \cos\theta}{\rho g r} \tag{2.3}$$

(*Jurin law*) where $g = 9.81\,\mathrm{m/s^2}$ is the acceleration of gravity, ρ is the density of the rising sap, γ is the surface tension (the up-lifting force caused by the liquid per unit of contact length inside a capillary tube), and θ is the so-called contact angle, *i.e.*, the angle that the line tangent to the liquid makes with the walls of the tube. This equation is obtained by balancing gravity with the vertical component of the capillary force (Fig. 2.7).

The weight of the column of liquid of height h in the capillary tube is

$$mg = \rho V g = \rho g h \pi r^2$$

where $V = \pi r^2 h$ is the volume of this column. The force supporting this column is the vertical component of the capillary force.

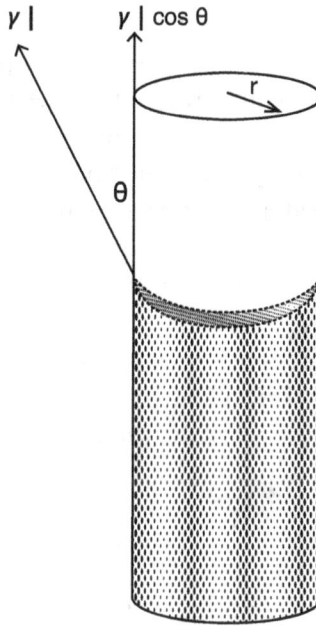

Fig. 2.7   A capillary tube.

The capillary force, tangent to the surface of the liquid, has intensity $\gamma l = 2\pi r \gamma$, where $l = 2\pi r$ is the length of contact between the liquid and the tube at the free surface. Given the contact angle θ, the vertical component of the capillary force is $\gamma l \cos \theta$ (Fig. 2.7). By balancing weight and capillary force in the vertical direction, we obtain

$$\rho g \pi r^2 h = \gamma \cdot 2\pi r \cos \theta,$$

which gives Eq. (2.3).

The question of whether sap can reach the top of a sequoia can now be answered. Assuming the sap density to be $\rho = 1.05 \cdot 10^3 \, \text{kg/m}^3$ (essentially that of water), its surface tension to be $\gamma = 7.28 \cdot 10^{-2} \, \text{N/m}$, $r \simeq 0.02 \, \text{mm}$, and the contact angle to be nearly zero, one obtains

$$h = \frac{2 \cdot \left(7.28 \cdot 10^{-2} \, \text{N/m}\right)}{(1.05 \cdot 10^3 \, \text{kg/m}^3) \cdot (9.81 \, \text{m/s}^2) \cdot (2 \cdot 10^{-5} \, \text{m})} = 0.71 \, \text{m},$$

more than a hundred times smaller than 80 m! It is clear that capillarity cannot win against gravity to raise sap even in small trees, let alone tall sequoias. Again, gravity rules. A molecular mechanism, instead, is responsible for raising sap in the xylem of trees. Water is lifted upward by both evapotranspiration at the top and the consequent motion of the water below that is drawn up in small tubular structures. The energy necessary to lift the water is provided by the latent heat released by evaporation from leaves located at the top of the tree.

Capillarity, however, works well horizontally, as I was reminded on my last hike when I accidentally left the sleeve of my sweater dipping in a pool of water next to where I was sitting.

Chapter 3

Water, snow, and ice

3.1 Introduction

Water is very important for the mountaineer, and not only to drink on a long climb. In its liquid form, it rushes from the heights making mountain streams that can be lovely to see and treacherous to cross, it runs on glaciers and below them, plunging from the surface to the depths through moulins. In its solid form, water makes snowfields and glaciers. Snow comes in a large variety of conditions, which make it difficult to judge the risk of avalanche on a slope, can make wonderful powder skiing, or reduce a snow climb to endless postholing. Ice also comes in various forms, from freshly formed waterfall ice so dear to ice climbers, to old black ice on alpine slopes at the end of the season, which is hard and difficult to keep crampon points in, to the vividly green ice exposed when a serac detaches from an ice wall and is best admired from far away. In its gaseous form, water builds clouds which engulf summits and climbers, bring rain, thunderstorms, or night dew, cut the visibility to nothing in unforgettable whiteouts, or make for dramatic scenery.

Water is present in all its forms in the mountains. Liquid water forms mountain streams, waterfalls, and glacial torrents which swell with melting. Solid water forms snowfields and glaciers which flow. Water vapour lifts and forms clouds and fog. Water often changes phase and even when it doesn't because of sub-zero temperatures, it still gives rise to interesting phenomena such as an enormous variety of shapes and types of snowflakes; flowing glaciers carving valleys and rocks and moving till and moraines; frozen waterfalls; seracs; the transport and deposition of snow by winds building snow pillows and cornices; and various types of avalanches. When the snow sublimates in very dry areas, it leaves behind curious formation of penitentes. In less dry areas, sublimation forms suncups instead, which seem to mirror effects and shapes encountered in experimental surface physics. There is much to see and understand for a naturalist.

3.2   Water is unusual

What is more common than water? 70% of the surface of the planet is covered by water and the oceans have an average depth of 3.8 km. Yet, from the scientific point of view, water has some unusual physical properties which set it apart from other common substances.

First, the boiling point of water is 100°C at 1 atmosphere of pressure, which is rather high. Second, the difference between melting point and boiling point is also rather large (100°C), and it is this feature that makes it possible to find water in the liquid state in nature and allows for the existence of oceans, lakes, and rivers on Earth. The fact that the temperature in the mountains easily drops below 0°C is responsible for the existence of large snowfields. The fact that, at higher elevations, the ground temperature stays below zero all year long or for most of the year makes it possible to have continued snowfalls and the accumulation of significant layers of snow. By compressing the layers underneath with its weight, this snow creates ice which begins to flow and form glaciers, and eventually turns onto ice itself.

Third, water *expands* when it turns into ice, unlike most common substances which contract when going from liquid to solid. As a consequence, ice does not sink to the bottom of lakes and oceans but it floats on top. It is a curious fact that the density of liquid water (*i.e.*, its mass per unit volume) increases as the temperature rises from 0°C, becoming maximum at 4°C and then decreases as the temperature rises. Hence, water contracts as the temperature rises between 0°C and 4°C. While a mass m of water stays constant, its volume V decreases and its density $\rho = m/V$ then increases. This is really bizarre in comparison with most substances which always expand (V increases) as the temperature rises. This fact allows life to continue under water, at least until the body of water freezes to the bottom, because denser water at 4°C sinks following Archimedes' law, while ice forms beginning from the top of a body of water and grows downward. Ice has insulating properties, so it protects deep water from the lower temperatures outside, especially if the ice gets covered by fresh snow which has a higher insulating power because of the air trapped in it (see Sec. 5.3). Another consequence of the peculiar property of water of expanding when turning from liquid to solid is that this expansion applies large stresses to the walls of a confining container, or to any confining boundary, which results in frost weathering, an important agent of erosion when the temperature oscillates frequently around the zero Celsius (Sec. 3.6).

The thermal properties of water are also unusual. The specific heat of water, that is, the heat energy necessary to raise the temperature of 1 kg of water by 1°C, is $4186 \, \mathrm{J} \cdot \mathrm{kg}^{-1} \cdot (°\mathrm{C})^{-1}$, which is unusually high among common materials. In fact, this numerical value is about five times that of the specific heat of most common rocks and soils. It is because of this relatively large number that cold water takes significantly longer to warm up than the surrounding rocks and soils and, conversely, warm water takes longer to cool down. Because of this effect, large bodies of water such as oceans and large lakes mitigate the climate with their thermal inertia and reduce temperature excursions. Green valleys filled with woods and other vegetation trap humidity, that is, water and water vapour and have a larger thermal inertia than the surroundings, which is the cause of mountain breezes (Sec. 5.6). The relatively high specific heat of water makes it also a good coolant for industrial applications, which may ultimately have some relevance in the manufacturing of our climbing gear but is otherwise irrelevant for the mountaineer.

Another peculiar feature of water is its latent heat of vaporization (the heat necessary to vaporize 1 kilogram of water at the pressure of 1 atmosphere): it is 2258 J/kg. This value is very large in comparison with most common substances, which means that a large amount of energy in the form of *latent heat* can be stored into water vapour. This energy was used to turn liquid into gas and is returned when this gas condenses into liquid again. For example, it is stored in water vapour evaporated at the tropics and residing in the lower atmosphere. This energy stored in water vapour is transported by global air circulation and is a very important item in the global distribution of solar energy (which caused the evaporation in the first place) around the planet. It is the latent heat of water vapour which fuels tropical hurricanes. When water vapour condenses, causing precipitation and continuing the hydrologic cycle, this energy is returned. Water is very efficient at putting out fires not only because it acts as a blanket depriving the flames of oxygen, but also because the liquid-vapour phase transition removes much of the heat and cools the fire much more than the evaporation of other liquids. The unusually high latent heat of water is also the reason why steam burns are worse than burns from liquid water. During the vapour-liquid phase transition, the steam condensing on skin releases more heat than liquid water would, and after this latent heat is released, we still have hot liquid water at 100°C. Watch out with camping pots!

Last but not least, an interesting physical-chemical property of water molecules is that they are polar, that is, they act like little electric dipoles, roughly speaking, little "sticks" with a positive electric charge at one end and a negative charge at the other end (Fig. 3.1). This is because the angle between the hydrogen bonds is 104.5°, in other words the molecule is not linear and the hydrogen atoms lie on one side. This makes negative charge reside preferentially near the oxygen atom and positive charges near the hydrogen atoms. The result is a net electric dipole.

Further, if another molecule with a dipole moment (solute) is present in a water solution, the water dipoles align with their negative ends surrounding the positive charges of the solute molecule, or *vice-versa*. This property makes water a very good solvent for many

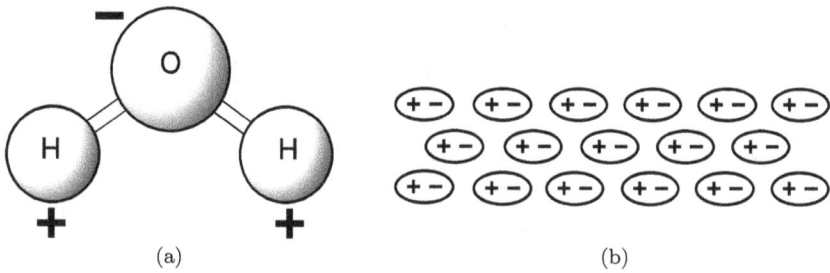

Fig. 3.1 (a) Because the hydrogen bonds make an angle of 104.5°, the H_2O molecule has some positive and some negative charges partially unbalanced and forms an electric dipole. (b) In liquid water these dipoles attract, forming weak bonds.

substances involved in biochemical reactions and a good medium to transport nutrients and waste products in plant and animal organisms. Don't forget to drink on your hike!

3.3 Skis, snowshoes, and pressure

Pressure is defined as

$$P = \frac{F}{A},$$

where F is the magnitude of the force applied on the area A *perpendicular* to it, when we picture the force (a vector) as an arrow (Fig. 3.2).

You can have a large force but, spread it over a large area A, and it results in a moderate pressure. *Vice-versa*, even a small force applied over a very small area results in a large pressure. When the wind pushes our tent, we want to reduce the normal area of the tent flap exposed to the wind, by going to a sheltered area, using a lower tent, or building a snow wall around the tent. When sailing, instead, we maximize the area A of the sail perpendicular to a moderate wind in order to maximize the wind force $F = PA$ on it.

The concept of pressure is applicable to skis: we sink in loose snow when we step on it with our boots because the vertical force of our weight is distributed over the relatively small area of the soles

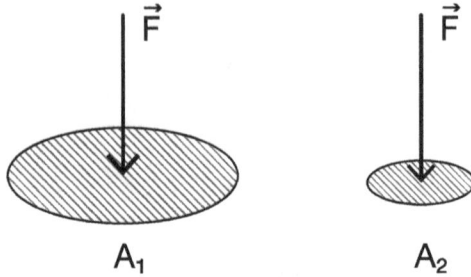

Fig. 3.2   Given equal force magnitude F, we have $P_1 = F/A_1 < P_2 = F/A_2$ because $A_2 < A_1$.

(say, $450\,\mathrm{cm}^2 = 0.045\,\mathrm{m}^2$ for my snow boots), resulting in a fairly large pressure. An 80 kg person will weight $F = mg = (80\,\mathrm{kg}) \cdot (9.8\,\mathrm{m/s}^2) = 780\,\mathrm{N}$, resulting in a pressure $P = F/A \simeq 17000\,\mathrm{Pa}$, more than enough to sink in powder snow. However, if this person puts his or her skis on, the same weight force is now distributed over a much larger area, say $A' = 0.26\,\mathrm{m}^2$ for my short alpine touring skis, resulting in the much smaller pressure $P' = F/A' = 3000\,\mathrm{Pa}$. Now the skis sink much less than the boots. If the same person uses snowshoes instead of skis (particularly appropriate with a thick cover of loose fluffy snow in terrain that is not steep, or in the forest where skis would get tangled in bushes all the time), he or she will sink even less because of the larger area of snowshoes, and the resulting lower pressure, in comparison with skis.

By the same principle, when one really can not avoid crossing a presumably thin snow bridge over a crevasse (Fig. 3.3), it is helpful to lie down and crawl over it, so as to spread the body weight over a larger surface and reduce the pressure on the snow.

Once, climbing Mt. Baker in the Pacific Northwest, I jumped a crevasse, landing in the spot where my two climbing partners lighter than me had already jumped. While the rather mushy snow on the other side held them okay, when I landed on the far side of the crevasse my leg sank all the way through the overhanging roof. Getting down immediately and spreading my weight over the snow greatly reduced the pressure applied to the snow and prevented me from breaking completely through it and falling down all the way,

Fig. 3.3 A snow bridge (Emerald Glacier, Canadian Rockies).

which was a good thing since I was the last one in the party and I had over ten meters of slack in the rope.

3.4 Under stress

There are forces acting within the snow and ice that we travel or that we look at from a distance. If we cross a snow bridge over a crevasse, we hope that these cohesive internal forces within the snow are strong enough to support our weight. Old snow sometimes builds bridges and structures which seem to defy equilibrium and are held together by internal stresses (Fig. 3.4).

Fig. 3.4 Internal stresses holding old snow in spring (Johnston Canyon, Canadian Rockies).

Fig. 3.5   Seracs on the Bionassay glacier (Mt. Blanc, France).

When we ski a slope that may not be completely safe (which should never happen), these internal stresses better hold the snow with the extra load of the skier on top. Similarly, seracs are held in place by internal stresses (Fig. 3.5) and if they aren't we see the result down below. These internal forces are not *bulk forces* like gravity, which is applied to every point inside the snow or ice: they are instead *surface forces*, applied across imaginary surfaces drawn within the snow or ice. Remember pressure? It was defined as a force divided by normal (*i.e.*, perpendicular) area. Stress is defined as force divided by *parallel* area. To make sense of this statement, imagine a parallelepiped of ice (or any material) sitting on a horizontal plane and a horizontal force of intensity F which tends to deform it (Fig. 3.6).

Let $A_\parallel$ be the area of the base of this ice block. The *shear stress* is the force per unit of parallel area applied to the side of the block

$$\sigma \equiv \frac{F}{A_\parallel}.$$

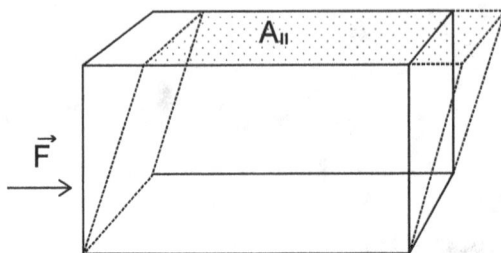

Fig. 3.6   Stress is force per unit of *parallel* area.

The shear stress is a quantity similar to pressure and has the same
dimensions and the same units (Pascals), except that it uses parallel
instead of perpendicular area and that, while the effect of pressure
is to compress the block, the effect of shear stress is to deform, or
shear, it. Similarly, one can define a stress associated with the other
horizontal direction and pushing the other side of the block. These
stresses are responsible for opposing the deformation of the material.
Think of glacier ice that is flowing subject to gravity and visualize
ice blocks inside of it. These blocks are kept together by the tensile
stresses acting across their various imaginary sides. When these tensile stresses fail, the blocks detach. This happens, for example, when
the weight of an overhanging serac pushed over an edge by the glacier
flow becomes too large for the tensile stresses to balance it and the
serac falls.

In general, one can consider any material which is not rigid and
define and study the stresses and pressures within it. This is done
in a branch of physics called *continuum mechanics*, which includes
fluids.[1] Yes, there are internal stresses, in addition to pressures, in
liquid water as well as in ice and, in retrospect, glacier ice flowing
downhill is just another fluid in slow motion.

There are internal stresses within a snowpack, too. When the
weight (a bulk force) acting on the snow overcomes these cohesive

---

[1]A *continuum* is a material which is not made of single point particles but is
distributed continuously over a region of space, for example a string in the one-
dimensional case, a plate in two dimensions, or a block of ice or a stream of water
in the 3-dimensional case.

stresses, we have an avalanche. This occurrence may be due to the fact that the ice crystals in the snowpack metamorphosize and the bonds between grains weaken (for example because of a temperature change), or it may be due to snowfall loading the slope further, or to the extra load due to the weight of a skier crossing it.

Similarly, cornices created by dominant winds during snowfall and/or by further transport of snow by the wind, protrude over mountain ridges and are supported only by internal stresses. Cornices break naturally and fall down, often on slopes or snow pillows loaded by snow transported onto the lee side of the mountain by the same wind that created the cornice. They often cause avalanches by falling on, and overloading, the slopes below them. Or they can fall because of mountaineers stepping on them, a very real danger when climbing ridge routes. There are many stories of mountaineers stepping over a cornice that gives way, and disappearing forever. Probably the most famous of such stories is that of the Austrian mountaineer Hermann Buhl, reputed to be one of the best climbers of all times, who died in 1957 by stepping on a cornice near the summit of Chogolisa in the Karakorum region of Pakistan. His body was never recovered.

The internal stresses appear in the equations ruling the motion or the static configuration of a material, solid or fluid. These equations relate external and internal forces (therefore, also stresses) with the acceleration of the material. Their form is rather complicated and solving these equations is even more complicated — usually they can only be solved by brute force on computers instead of using arrays of analytical techniques which are known for the simpler ordinary differential equations [Hille (1969); Brauer and Noel (1986)] but do not work for a continuous medium. The response of materials to stresses vary greatly: for example an elastic solid will display a shear stress that is proportional to the deformation it is subject to, but ice and snow are not so simple and the relation between deformation (or velocities) and stresses is non-linear, which is one of the sources of the complication of the equations describing the motion or even the static configuration of these materials.

For once, we now break from the style followed thus far and we provide a technical discussion of what already said in this section to give the flavour of a precise mathematical description to the readers who may wonder about it. Warning: this discussion requires linear algebra [Lang (1987)]. The reader who is not familiar with this area of mathematics and does not desire more precision can safely skip to the next section without missing any new physical concept.

Focus on contact forces which act on surfaces within the material. Consider a point P in glacier ice, or in snow, or in any material (for simplicity we refer to glacier ice, but what said holds true for most materials). Consider also an arbitrary direction through this point P identified by a unit vector $\vec{n}$. Further consider a real or imaginary surface through the point P and an element of plane perpendicular to the direction $\vec{n}$ at P and with area δA.

The material on the side of this plane which lies in the direction of $\vec{n}$ ("positive side", denoted with "+") exerts a force $\delta \vec{F}$ on the material on the other side ("negative side", denoted with a "−") through this surface. The *stress at the point P across the plane perpendicular to the direction $\vec{n}$* is the limit

$$\vec{S}(\vec{n}) = \lim_{\delta A \to 0} \frac{\delta \vec{F}}{\delta A}. \tag{3.1}$$

It is clear from this definition that a stress is a force per unit area, and then it has the dimensions of a pressure. Stresses in glaciers have typical magnitudes in the range 50–200 kPa.

The material on the positive side of δA exerts a force $\vec{S}\delta A$ on the material on the negative side, while the latter exerts a force $-\vec{S}\delta A$ on the positive side. For each arbitrary direction $\vec{n}$ through P there is a stress $\vec{S}(\vec{n})$. We are dealing with two vectors at once: one describing the arbitrary direction $\vec{n}$ at P and one describing the stress $\vec{S}(\vec{n})$ once the direction $\vec{n}$ is assigned. In general, the stress $\vec{S}$ is not parallel to the normal $\vec{n}$, but it is directed sideways. As the direction $\vec{n}$ varies, and an index is assigned to its components n^i, the stress $\vec{S}(\vec{n})$ will also vary, and an index is assigned to its components S^j. Therefore, we need an object with two indices, which is a matrix or a 2-tensor [Spain (2003)].

For illustration, consider first the special situation in which the direction $\vec{n}$ at P is along the x-axis (Fig. 3.7, we will generalize later).

Then the plane perpendicular to $\vec{n}$ at the point P is the (y, z) plane and the stress $\vec{S}$ at P is described by its Cartesian components

$$\vec{S}(\vec{n} = \vec{e}_x) = (\sigma_{xx}, \sigma_{xy}, \sigma_{xz}). \tag{3.2}$$

The component σ_{xx} of $\vec{S}$ in the $\vec{n}$ direction *perpendicular to* δA is called *normal stress*. Being a force per unit of normal area, it is a pressure or a tension, depending on its sign. By convention a *tension* is a positive normal stress, while a *compression* is a negative normal stress.

The components σ_{xy} and σ_{xz} of $\vec{S}$ are parallel to the surface δA and are called *shear stresses*. Contrary to pressures, which are forces per unit of *normal* area, shear stresses are forces per unit of *parallel* area.

The component σ_{xy} is perpendicular to the x-axis and parallel to the y-axis, while the component σ_{xz} is perpendicular to the x-axis and parallel to the z-axis.

Suppose now that the $\vec{n}$ direction is aligned along the y-axis: then the (x, z) plane is perpendicular to $\vec{n}$ and the stress $\vec{S}$ at P is described by its components

$$\vec{S}(\vec{n} = \vec{e}_y) = (\sigma_{yx}, \sigma_{yy}, \sigma_{yz}). \tag{3.3}$$

The σ_{yx} component is perpendicular to the y-axis and parallel to the x-axis, while the σ_{yy} component is directed along the y-axis and the σ_{yz} component is perpendicular to the y-axis and parallel to the z-axis.

If, instead, we assume that the $\vec{n}$ direction is along the z-axis, then the stress at P is

$$\vec{S}(\vec{n} = \vec{e}_z) = (\sigma_{zx}, \sigma_{zy}, \sigma_{zz}). \tag{3.4}$$

The σ_{zx} component is perpendicular to the z-axis and parallel to the x-axis, while the σ_{zy} component is perpendicular to the z-axis and parallel to the y-axis and the σ_{zz} component is directed along the z-axis.

Now let us generalize. In general, the direction $\vec{n} = (n^x, n^y, n^z)$ is not directed along any of the Cartesian axes and there are

9 components which form the *stress tensor* described, in three spatial dimensions, by the 3×3 matrix

$$\hat{\sigma} = (\sigma_{ij}) = \begin{pmatrix} \sigma_{xx} & \sigma_{xy} & \sigma_{xz} \\ \sigma_{yx} & \sigma_{yy} & \sigma_{yz} \\ \sigma_{zx} & \sigma_{zy} & \sigma_{zz} \end{pmatrix} \qquad (3.5)$$

$(i, j = 1, 2, 3)$. The elements σ_{ii} of the main diagonal of the stress tensor are normal stresses (compressions or tensions), while the off-diagonal elements σ_{ij} with $j \neq i$ are shear stresses. The component σ_{ii} is directed along the i-th axis, while the component σ_{ij} with $i \neq j$ is perpendicular to the i-th axis and parallel to the j-th axis. Fig. 3.7 summarizes the components of the stress tensor.

A fundamental property of the stress tensor is that it is symmetric, $\sigma_{ji} = \sigma_{ij}$ for all $i, j = 1, 2, 3$. It can be shown that this property is necessary for the rotational equilibrium of any cube considered inside the material [Paterson (1994); Hooke (2005)].

Because it is symmetric,[2] the tensor σ_{ij} in 3 spatial dimensions has at most 6 independent components and can be written as

$$\hat{\sigma} = (\sigma_{ij}) = \begin{pmatrix} \sigma_{xx} & \sigma_{xy} & \sigma_{xz} \\ \sigma_{xy} & \sigma_{yy} & \sigma_{yz} \\ \sigma_{xz} & \sigma_{yz} & \sigma_{zz} \end{pmatrix}. \qquad (3.6)$$

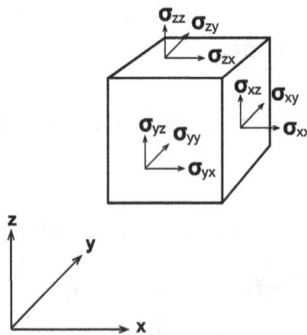

Fig. 3.7   The stresses acting on a cube inside a continuous medium.

---

[2]In general, a symmetric 2-index tensor in n spatial dimensions has at most $n(n+1)/2$ independent components.

The stress vector at a point P perpendicular to the direction $\vec{n}$ has components[3]

$$S_i(\vec{n}) = \sum_{j=1}^{3} \sigma_{ij} n^j \tag{3.7}$$

or, in matrix notation,

$$
\vec{S}(\vec{n}) = \begin{pmatrix} \sigma_{xx} & \sigma_{xy} & \sigma_{xz} \\ \sigma_{xy} & \sigma_{yy} & \sigma_{yz} \\ \sigma_{xz} & \sigma_{yz} & \sigma_{zz} \end{pmatrix} \begin{pmatrix} n_x \\ n_y \\ n_z \end{pmatrix} = \begin{pmatrix} \sigma_{xx} n_x + \sigma_{xy} n_y + \sigma_{xz} n_z \\ \sigma_{xy} n_x + \sigma_{yy} n_y + \sigma_{yz} n_z \\ \sigma_{xz} n_x + \sigma_{yz} n_y + \sigma_{zz} n_z \end{pmatrix}
$$

$$
= \begin{pmatrix} S_x \\ S_y \\ S_z \end{pmatrix} \tag{3.8}
$$

while, in dyadic notation [Goldstein (1980)], we have

$$\vec{S}(\vec{n}) = \vec{\vec{\sigma}} \cdot \vec{n}. \tag{3.9}$$

At every point in the material one can compute the arithmetic average of the normal stresses, which is called *pressure*,

$$P \equiv \frac{\sigma_{xx} + \sigma_{yy} + \sigma_{zz}}{3} = \frac{\mathrm{Tr}(\sigma_{ij})}{3}. \tag{3.10}$$

P is close, but not necessarily equal, to the cryostatic pressure. The *deviatoric stress* or *stress deviator* is the tensor

$$s_{ij} \equiv \sigma_{ij} - P\delta_{ij} = \sigma_{ij} - \frac{\mathrm{Tr}(\hat{\sigma})}{3} \delta_{ij}, \tag{3.11}$$

which is nothing but the trace-free part of the stress tensor describing the deviation of the stresses from the average (3.10). To understand this definition, it is useful to recall a general property of tensors.

---

[3]In this section we use only Cartesian coordinates, in which covariant and contravariant components of a vector coincide and the position of the indices (high or low) does not matter [Spain (2003); Faraoni (2013)].

A 2-index tensor T_{ij} which, in general, is not symmetric nor anti-symmetric in its indices, can be decomposed into its *symmetric part* S_{ij} and its *antisymmetric part* A_{ij} as follows [Spain (2003)]:

$$T_{ij} = \frac{T_{ij} + T_{ji}}{2} + \frac{T_{ij} - T_{ji}}{2} \equiv S_{ij} + A_{ij}. \tag{3.12}$$

This identity is trivial to verify and it can be shown easily that the decomposition is unique [Faraoni (2006)]. The symmetric part $S_{ij} \equiv (T_{ij} + T_{ji})/2$ can be further decomposed into a trace-free part and a pure trace part,

$$S_{ij} = \left[S_{ij} - \frac{\mathrm{Tr}(\hat{\mathbf{S}})}{3} \delta_{ij} \right] + \frac{\mathrm{Tr}(\hat{\mathbf{S}})}{3} \delta_{ij} \equiv \bar{S}_{ij} + \frac{\mathrm{Tr}(\hat{\mathbf{S}})}{3} \delta_{ij}. \tag{3.13}$$

The tensor $\bar{S}_{ij}$ has zero trace by construction, while $\frac{\mathrm{Tr}(\hat{\mathbf{S}})}{3} \delta_{ij}$ is built out of the trace of S_{ij}, which coincides with the trace of T_{ij} since the antisymmetric part A_{ij} of T_{ij} has zero trace, like all antisymmetric tensors, for which all the diagonal components vanish by virtue of the antisymmetry, since $A_{ji} = -A_{ij}$ implies $A_{ii} = -A_{ii} = 0$.

The independent components of the deviatoric stress tensor are

$$s_{xx} = \sigma_{xx} - P = \frac{2\sigma_{xx} - \sigma_{yy} - \sigma_{zz}}{3}, \tag{3.14}$$

$$s_{yy} = \sigma_{yy} - P = \frac{2\sigma_{yy} - \sigma_{xx} - \sigma_{zz}}{3}, \tag{3.15}$$

$$s_{zz} = \sigma_{zz} - P = \frac{2\sigma_{zz} - \sigma_{yy} - \sigma_{xx}}{3}, \tag{3.16}$$

$$s_{xy} = \sigma_{xy}, \tag{3.17}$$

$$s_{yz} = \sigma_{yz}, \tag{3.18}$$

$$s_{xz} = \sigma_{xz}. \tag{3.19}$$

The off-diagonal components coincide with the components of the stress tensor. The diagonal components of the deviatoric stress are called *stress deviators* and their average vanishes:

$$\frac{s_{xx} + s_{yy} + s_{zz}}{3} = \frac{(\sigma_{xx} - P) + (\sigma_{yy} - P) + (\sigma_{zz} - P)}{3}$$

$$= \frac{\sigma_{xx} + \sigma_{yy} + \sigma_{zz}}{3} - P \equiv P - P = 0. \quad (3.20)$$

As for any 2-index tensor or matrix, the stress tensor σ_{ij} can be decomposed into a pure trace and a trace-free part:

$$\sigma_{ij} = \left[\sigma_{ij} - \frac{\text{Tr}(\hat{\sigma})}{3} \delta_{ij} \right] + \frac{\text{Tr}(\hat{\sigma})}{3} \delta_{ij} \equiv s_{ij} + \frac{\text{Tr}(\hat{\sigma})}{3} \delta_{ij}. \quad (3.21)$$

The statement (3.20) that the average of the deviatoric stresses s_{ii} vanishes is then nothing but the statement that this quantity is the trace of the deviatoric stress tensor s_{ij} which, by construction, is trace-free: $\text{Tr}(s_{ij}) = 0$. Being diagonal, the pure trace part of σ_{ij} contributes only to the normal stresses but not to the shear stresses.

$$\hat{s} = (s_{ij}) = \begin{pmatrix} \sigma_{xx} - P & \sigma_{xy} & \sigma_{xz} \\ \sigma_{xy} & \sigma_{yy} - P & \sigma_{xz} \\ \sigma_{xz} & \sigma_{yz} & \sigma_{zz} - P \end{pmatrix} \quad (3.22)$$

$$= (\sigma_{ij}) - \begin{pmatrix} P & 0 & 0 \\ 0 & P & 0 \\ 0 & 0 & P \end{pmatrix}. \quad (3.23)$$

We can construct now another invariant of the deviatoric stress tensor $\hat{s}$, in addition to its trace $\text{Tr}(\hat{s})$ and its determinant $\text{Det}(\hat{s})$, by using the tensor

$$\hat{s}^2 = \begin{pmatrix} s_{xx} & s_{yy} & s_{xz} \\ s_{xy} & s_{yy} & s_{yz} \\ s_{xz} & s_{yz} & s_{zz} \end{pmatrix} \begin{pmatrix} s_{xx} & s_{yy} & s_{xz} \\ s_{xy} & s_{yy} & s_{yz} \\ s_{xz} & s_{yz} & s_{zz} \end{pmatrix} =$$

$$\begin{pmatrix} s_{xx}^2 + s_{xy}^2 + s_{xz}^2 & s_{xx}s_{xy} + s_{xy}s_{yy} + s_{xz}s_{yz} & s_{xx}s_{xz} + s_{xy}s_{yz} + s_{xz}s_{zz} \\ s_{xy}s_{xx} + s_{xy}s_{yy} + s_{yz}s_{xz} & s_{xy}^2 + s_{yy}^2 + s_{yz}^2 & s_{xz}s_{xy} + s_{yy}s_{yz} + s_{zz}s_{yz} \\ s_{xx}s_{xz} + s_{xy}s_{yz} + s_{zz}s_{xz} & s_{xy}s_{xz} + s_{yy}s_{yz} + s_{zz}s_{yz} & s_{xz}^2 + s_{yz}^2 + s_{zz}^2 \end{pmatrix},$$

$$(3.24)$$

which has trace

$$
\begin{aligned}
\mathrm{Tr}(\hat{s}^2) &= \left(s_{xx}^2 + s_{xy}^2 + s_{xz}^2\right) + \left(s_{xy}^2 + s_{yy}^2 + s_{yz}^2\right) + \left(s_{xz}^2 + s_{yz}^2 + s_{zz}^2\right) \\
&= s_{xx}^2 + s_{yy}^2 + s_{zz}^2 + 2\left(s_{xy}^2 + s_{yz}^2 + s_{xz}^2\right) \\
&= \left[(s_{xx} + s_{yy} + s_{zz})^2 - 2s_{xx}s_{yy} - 2s_{yy}s_{zz} - 2s_{xx}s_{zz}\right] \\
&\quad + 2\left(s_{xy}^2 + s_{yz}^2 + s_{xz}^2\right) \\
&= 2\left[s_{xy}^2 + s_{yz}^2 + s_{xz}^2 - (s_{xx}s_{yy} + s_{yy}s_{zz} + s_{xx}s_{zz})\right]. \quad (3.25)
\end{aligned}
$$

The *effective stress* is defined as

$$
s_{\mathrm{eff}} \equiv \sqrt{\frac{1}{2}\sum_{i,j=1}^{3} s_{ij}^2} = \sqrt{\frac{1}{2}\mathrm{Tr}(\hat{s}^2)} \quad (3.26)
$$

$$
= \sqrt{s_{xy}^2 + s_{yz}^2 + s_{xz}^2 - (s_{xx}s_{yy} + s_{yy}s_{zz} + s_{xx}s_{zz})} \quad (3.27)
$$

$$
= \frac{1}{\sqrt{2}}\sqrt{s_{xx}^2 + s_{yy}^2 + s_{zz}^2 + 2\left(s_{xy}^2 + s_{yz^2} + s_{xz}^2\right)}. \quad (3.28)
$$

The last equality is proved in Appendix D.

The effective stress is a sort of "norm of the matrix $\hat{s}$" and is an invariant associated with $\hat{s}$. The effective stress depends on all the components of the deviatoric stress. Now, theory and experiment suggest that the strain rates in ice and the ice velocity in a given direction depend on *all* the components of the deviatoric stress tensor involving also the other directions, not only those in the given direction. This feature is reflected in the use of the effective stress in glaciology.

Similar to the effective stress, one defines also the quantity σ_{eff} using σ_{ij} instead of s_{ij},

$$
\sigma_{\mathrm{eff}} \equiv \sqrt{\frac{1}{2}\sum_{i,j=1}^{3} \sigma_{ij}^2} = \sqrt{\frac{1}{2}\mathrm{Tr}(\hat{\sigma}^2)}. \quad (3.29)
$$

Then the relation between $s_{\mathrm{eff}}, \sigma_{\mathrm{eff}}$, and P holds:

$$
s_{\mathrm{eff}}^2 = \sigma_{\mathrm{eff}}^2 + 3P^2. \quad (3.30)
$$

Consider now, in two spatial dimensions for simplicity, an infinitesimal element of material shaped like a parallelepiped with sides of lengths $\delta x, \delta y$, and δz oriented along the coordinate axes. The face in the (x, y) plane has area $\delta x \delta y$. The net force per unit length in the x-direction acting on the parallelepiped is

$$\left[\left(\sigma_{xx} + \frac{\partial \sigma_{xx}}{\partial x} \delta x \right) - \sigma_{xx} \right] \delta y\, \delta z + \left[\left(\sigma_{xy} + \frac{\partial \sigma_{xy}}{\partial y} \delta y \right) - \sigma_{xy} \right] \delta x\, \delta z$$
$$= \left(\frac{\partial \sigma_{xx}}{\partial x} + \frac{\partial \sigma_{xy}}{\partial y} \right) \delta x\, \delta y\, \delta z. \tag{3.31}$$

These forces act on the surface $\delta x \delta y$ of the material element, while gravity (and, in principle, other bulk forces) acts on all points of the material. Let $\vec{f} = (f^x, f^y, f^z)$ be the (total) bulk force per unit mass in the material (for glacier ice, this is only gravity, which is vertical) and let ρ be the mass density of the material. Then, the x-component of the total bulk force is $\rho f^x \delta x\, \delta y\, \delta z$. The acceleration of the material element is $\vec{a} = (a^x, a^y, a^z) = \ddot{\vec{\xi}}$, where $\vec{\xi}$ is the displacement of a point in the material (see below) and an overdot denotes differentiation with respect to time. Newton's second law of motion is then written as

$$\frac{\partial \sigma_{xx}}{\partial x} + \frac{\partial \sigma_{xy}}{\partial y} + \rho f_x = \rho \ddot{\xi}_x, \tag{3.32}$$

$$\frac{\partial \sigma_{yy}}{\partial y} + \frac{\partial \sigma_{yx}}{\partial x} + \rho f_y = \rho \ddot{\xi}_y. \tag{3.33}$$

These are the mechanical equations of motion of glacier ice. In three dimensions they generalize to

$$\frac{\partial \sigma_{xx}}{\partial x} + \frac{\partial \sigma_{xy}}{\partial y} + \frac{\partial \sigma_{xz}}{\partial z} + \rho f_x = \rho \ddot{\xi}_x, \tag{3.34}$$

$$\frac{\partial \sigma_{yy}}{\partial y} + \frac{\partial \sigma_{yz}}{\partial z} + \frac{\partial \sigma_{yx}}{\partial x} + \rho f_y = \rho \ddot{\xi}_y, \tag{3.35}$$

$$\frac{\partial \sigma_{zz}}{\partial z} + \frac{\partial \sigma_{zx}}{\partial x} + \frac{\partial \sigma_{zy}}{\partial y} + \rho f_z = \rho \ddot{\xi}_z. \tag{3.36}$$

In compact form, we have

$$\sum_{j=1}^{3} \frac{\partial \sigma_{ij}}{\partial x^j} + \rho f_i = \rho \ddot{\xi}_i. \tag{3.37}$$

When the acceleration of the material element vanishes, as is customary for models of glaciers in steady state, $\ddot{\xi}_i = 0$, Eqs. (3.34)–(3.36) reduce to the *stress-equilibrium* (or *momentum balance*) *equations*

$$\sum_{j=1}^{3} \frac{\partial \sigma_{ij}}{\partial x^j} + \rho f_i = 0. \tag{3.38}$$

Let us introduce strains. Under the action of stresses, a material is deformed and its points are displaced from the positions that they would occupy in the absence of stresses. Let the vector $\vec{\xi}$ be the displacement of a point $\vec{x}$. Since the material is not rigid, the displacement of a point in its interior depends on its position, $\vec{\xi} = \vec{\xi}(\vec{x})$. Consider the displacements of a point at position $\vec{x}$ and of a nearby point at $\vec{x} + \delta\vec{x}$:

$$\vec{x} \longrightarrow \vec{x} + \vec{\xi}(\vec{x}), \tag{3.39}$$

$$\vec{x} + \delta\vec{x} \longrightarrow \vec{x} + \vec{\xi}(\vec{x} + \delta\vec{x}), \tag{3.40}$$

or, in components,

$$x^i \longrightarrow x^i + \xi^i(\vec{x}), \tag{3.41}$$

$$x^i + \delta x^i \longrightarrow x^i + \xi^i(\vec{x} + \delta\vec{x}) = x^i + \xi^i(\vec{x}) + \sum_{j=1}^{3} \frac{\partial \xi^i}{\partial x^j} \delta x^j + \cdots \tag{3.42}$$

Assume that the displacements are small and retain only the first order in the expansion (3.42). The gradient of the displacement vector $\vec{\nabla}\vec{\xi} \equiv \frac{\partial \vec{\xi}}{\partial \vec{x}}$ then appears in our equations. Its components are $\partial \xi^i / \partial x^j$ ($i, j = 1, 2, 3$). Like any 2-index tensor (or matrix), the displacement gradient can be decomposed into a symmetric and an antisymmetric part as follows:

$$\frac{\partial \xi^i}{\partial x^j} = \frac{1}{2}\left(\frac{\partial \xi^i}{\partial x^j} + \frac{\partial \xi^j}{\partial x^i}\right) + \frac{1}{2}\left(\frac{\partial \xi^i}{\partial x^j} - \frac{\partial \xi^j}{\partial x^i}\right) \equiv \epsilon_{ij} + \omega_{ij} \qquad (3.43)$$

This identity is trivial to verify. It is easy to show that this tensor decomposition into symmetric and antisymmetric parts is unique [Faraoni (2006)]. One can further decompose the symmetric part into a trace-free part and a pure trace part

$$\epsilon_{ij} = \left[\epsilon_{ij} - \left(\frac{\sum_{k=1}^{3}\epsilon_{kk}}{3}\right)\delta_{ij}\right] + \left(\frac{\sum_{k=1}^{3}\epsilon_{kk}}{3}\right)\delta_{ij}$$

$$\equiv \bar{\epsilon}_{ij} + \left(\frac{\sum_{k=1}^{3}\epsilon_{kk}}{3}\right)\delta_{ij}. \qquad (3.44)$$

The tensor $\bar{\epsilon}_{ij}$ is trace-free,

$$\sum_{k=1}^{3}\bar{\epsilon}_{kk} = \sum_{k=1}^{3}\left[\epsilon_{kk} - \frac{\sum_{k=1}^{3}\epsilon_{kk}}{3}\delta_{kk}\right] = \sum_{k=1}^{3}\epsilon_{kk} - \frac{\sum_{k=1}^{3}\epsilon_{kk}}{3}\cdot 3 = 0$$

$$(3.45)$$

and

$$\frac{1}{3}\sum_{k=1}^{3}\epsilon_{kk} = \frac{1}{3}\sum_{k=1}^{3}\frac{\partial \xi^k}{\partial x^k} = \frac{1}{3}\vec{\nabla}\cdot\vec{\xi}, \qquad (3.46)$$

so that

$$\epsilon_{ij} = \bar{\epsilon}_{ij} + \frac{\vec{\nabla}\cdot\vec{\xi}}{3}\delta_{ij}. \qquad (3.47)$$

ϵ_{ij} is the *strain tensor*. In Cartesian coordinates, it is dimensionless, $[\epsilon_{ij}] = [0]$. Its components are

$$\epsilon_{xx} = \frac{1}{2}\left(\frac{\partial \xi^x}{\partial x} + \frac{\partial \xi^x}{\partial x}\right) = \frac{\partial \xi^x}{\partial x}, \qquad (3.48)$$

$$\epsilon_{xy} = \frac{1}{2}\left(\frac{\partial \xi^x}{\partial y} + \frac{\partial \xi^y}{\partial x}\right) = \epsilon_{yx}, \qquad (3.49)$$

$$\epsilon_{xz} = \frac{1}{2}\left(\frac{\partial \xi^x}{\partial z} + \frac{\partial \xi^z}{\partial x}\right) = \epsilon_{zx}, \tag{3.50}$$

$$\epsilon_{yy} = \frac{1}{2}\left(\frac{\partial \xi^y}{\partial y} + \frac{\partial \xi^y}{\partial y}\right) = \frac{\partial \xi^y}{\partial y}, \tag{3.51}$$

$$\epsilon_{yz} = \frac{1}{2}\left(\frac{\partial \xi^y}{\partial z} + \frac{\partial \xi^z}{\partial y}\right) = \epsilon_{zy}, \tag{3.52}$$

$$\epsilon_{zz} = \frac{1}{2}\left(\frac{\partial \xi^z}{\partial z} + \frac{\partial \xi^z}{\partial z}\right) = \frac{\partial \xi^z}{\partial z}. \tag{3.53}$$

The strain tensor is, by definition, symmetric

$$\epsilon_{ji} = \epsilon_{ij} \tag{3.54}$$

and is represented by the matrix

$$\hat{\epsilon} = (\epsilon_{ij}) = \begin{pmatrix} \epsilon_{xx} & \epsilon_{xy} & \epsilon_{xz} \\ \epsilon_{xy} & \epsilon_{yy} & \epsilon_{yz} \\ \epsilon_{xz} & \epsilon_{yz} & \epsilon_{zz} \end{pmatrix}. \tag{3.55}$$

The diagonal elements ϵ_{ii} are called *normal strains* while the off-diagonal elements ϵ_{ij} with $j \neq i$ are called *shear strains*. The anti-symmetric part ω_{ij} of the displacement gradient (sometimes called *rotation tensor*) has components

$$\omega_{xx} = \omega_{yy} = \omega_{zz} = 0, \tag{3.56}$$

$$\omega_{xy} = \frac{1}{2}\left(\frac{\partial \xi^x}{\partial y} - \frac{\partial \xi^y}{\partial x}\right) = -\omega_{yx}, \tag{3.57}$$

$$\omega_{xz} = \frac{1}{2}\left(\frac{\partial \xi^x}{\partial z} - \frac{\partial \xi^z}{\partial x}\right) = -\omega_{zx}, \tag{3.58}$$

$$\omega_{yz} = \frac{1}{2}\left(\frac{\partial \xi^y}{\partial z} - \frac{\partial \xi^z}{\partial y}\right) = -\omega_{zy}. \tag{3.59}$$

The time derivative of the displacement $\vec{\xi}$ is identified with the velocity of the ice, $\dot{\vec{\xi}} = \vec{v}$ when the glacier moves solely because of the ice creep and not because of sliding along its bed or of sub-glacial till deformation [Paterson (1994); Cuffey and Paterson (2010); Hooke (2005)]. Then the spatial gradient of the velocity $\vec{\nabla}\vec{v} = \frac{\partial \vec{v}}{\partial \vec{x}}$ is described by the *strain rate* tensor

$$\hat{\dot{\epsilon}} = (\dot{\epsilon}_{ij}) = \begin{pmatrix} \dot{\epsilon}_{xx} & \dot{\epsilon}_{xy} & \dot{\epsilon}_{xz} \\ \dot{\epsilon}_{xy} & \dot{\epsilon}_{yy} & \dot{\epsilon}_{yz} \\ \dot{\epsilon}_{xz} & \dot{\epsilon}_{yz} & \dot{\epsilon}_{zz} \end{pmatrix}, \tag{3.60}$$

which is more useful than the strain tensor itself in the description of glacier ice deformations in time.

In Cartesian coordinates the strain tensor has the dimensions of the inverse of a time, $[\dot{\epsilon}_{ij}] = [T^{-1}]$, and its components are

$$\dot{\epsilon}_{ij} \equiv \frac{\partial \epsilon_{ij}}{\partial t} = \frac{1}{2} \frac{\partial}{\partial t} \left(\frac{\partial \xi^i}{\partial x^j} + \frac{\partial \xi^j}{\partial x^i} \right) = \frac{1}{2} \left(\frac{\partial \dot{\xi}^i}{\partial x^j} + \frac{\partial \dot{\xi}^j}{\partial x^i} \right). \tag{3.61}$$

A material is described by a constitutive relation, which is an equation linking stresses and strains, or stresses and strain rates.

The trace of the strain rate tensor is the divergence of the velocity $\vec{v}$ of the ice flow:

$$\text{Tr}\,(\dot{\epsilon}_{ij}) = \dot{\epsilon}_{xx} + \dot{\epsilon}_{yy} + \dot{\epsilon}_{zz} = \frac{\partial \dot{\xi}^x}{\partial x} + \frac{\partial \dot{\xi}^y}{\partial y} + \frac{\partial \dot{\xi}^z}{\partial z} = \vec{\nabla} \cdot \vec{v}. \tag{3.62}$$

In the absence of sources or sinks, the velocity of a fluid is related to its density ρ by the continuity equation

$$\frac{\partial \rho}{\partial t} + \vec{\nabla} \cdot (\rho \vec{v}) = 0 \tag{3.63}$$

expressing mass conservation. Since ice is nearly incompressible, the density ρ is assumed to be constant and it follows that

$$\vec{\nabla} \cdot \vec{v} = \dot{\epsilon}_{xx} + \dot{\epsilon}_{yy} + \dot{\epsilon}_{zz} = 0. \tag{3.64}$$

Similar to the effective stress, the *effective strain rate* is

$$\dot{\epsilon}_{\text{eff}} \equiv \sqrt{\frac{1}{2} \sum_{i,j=1}^{3} \dot{\epsilon}_{ij}^2} = \sqrt{\frac{1}{2} \text{Tr}(\hat{\dot{\epsilon}}^2)} \tag{3.65}$$

$$= \sqrt{\dot{\epsilon}_{xy}^2 + \dot{\epsilon}_{yz}^2 + \dot{\epsilon}_{xz}^2 - (\dot{\epsilon}_{xx}\dot{\epsilon}_{yy} + \dot{\epsilon}_{yy}\dot{\epsilon}_{zz} + \dot{\epsilon}_{xx}\dot{\epsilon}_{zz})} \tag{3.66}$$

$$= \frac{1}{\sqrt{2}} \sqrt{\dot{\epsilon}_{xx}^2 + \dot{\epsilon}_{yy}^2 + \dot{\epsilon}_{zz}^2 + 2\left(\dot{\epsilon}_{xy}^2 + \dot{\epsilon}_{yz^2} + \dot{\epsilon}_{xz}^2\right)} \tag{3.67}$$

and it is an invariant of the tensor $\dot{\epsilon}_{ij}$.

Assuming glacier ice to be isotropic and incompressible, its constitutive relation is given by[4] the *Glen law* [Glen (1952, 1955)]

$$\dot{\epsilon}_{\text{eff}} = A\, s_{\text{eff}}^n, \tag{3.68}$$

where A is a constant that depends on the temperature, crystal orientation, and on the presence of impurities in the ice, with dimensions

$$[A] = \left[\frac{\text{Pa}^{-n}}{\text{s}}\right]. \tag{3.69}$$

$n = 3$ describes ice flow. In isotropic ice, the Glen law (3.68) can be written in the form

$$\dot{\epsilon}_{ij} = A\, s_{\text{eff}}^{n-1}\, s_{ij}. \tag{3.70}$$

Slightly different forms of the Glen law are found in the literature. Sometimes Eq. (3.68) is written as

$$\dot{\epsilon}_{\text{eff}} = \left(\frac{s_{\text{eff}}}{B}\right)^n, \tag{3.71}$$

where $B = A^{-1/n}$ is a sort of viscosity parameter with dimensions

$$[B] = \left[\text{Pa} \cdot \text{s}^{1/n}\right] \tag{3.72}$$

(although glaciologists would use as units megapascals and years instead of SI units). In the system of principal axes in which the deviatoric stress and the strain rate tensors are diagonal, Eq. (3.70) becomes

$$\dot{\epsilon}_{ij} = \frac{s_{\text{eff}}^{n-1}}{B^n}\, s_{ij}. \tag{3.73}$$

The constant A depends on the temperature T according to the generic Arrhenius relation[5]

$$A(T) = A_0\, e^{-\frac{Q}{RT}}, \tag{3.74}$$

---

[4] This law was already suggested by Max Perutz, then an ice expert who later turned to biochemistry and won the 1962 Nobel Prize for Chemistry [Perutz (1950)].

[5] The Arrhenius relation is well known in chemistry and rules a variety of phenomena including the rates of chemical reactions and gives, for example, the temperature dependence of diffusion coefficients and the population of crystal vacancies.

where $\mathcal{A}_0$ is a (temperature-independent) constant, $R = 8.314 \cdot 10^3$ J $\cdot$ kmol$^{-1} \cdot$ k is the universal gas constant, Q is the *activation energy for creep*, and T is measured in Kelvins. Laboratory experiments give 42 kJ/mol $\leq Q \leq 84$ kJ/mol for temperatures $T < -10°$C and $Q \simeq 139$ kJ/mol for $T > -10°$C.

If a glacier is not isothermal, then $\mathcal{A} = \mathcal{A}(T)$ cannot be treated as a constant because it depends on the position and then the ice is not homogeneous.

3.5 Phase changes

We are all familiar with some changes of phase: common phase transitions are changes from the solid to the liquid state (melting), or from the liquid to the gas state (evaporation), and *vice-versa* (freezing and condensation, respectively). Water vapour in the air condenses on cold surfaces (*cold wall principle*), which explains the morning fog which forms at night when the ground temperature drops, and lingers for a few hours in the morning (Figs. 3.8 and 7.2). Another manifestation of phase transitions is *verglas*, very thin ice forming on the surface of rocks from the solidification of water vapour in the air and too thin for crampon points to bite, which makes alpine climbs extremely treacherous.

From the microscopic point of view, the atoms or molecules of a gas have maximum freedom to run away from each other in the three spatial dimensions, hitting each other every now and then, or possibly hitting the walls of a container. These repeated and frequent collisions with the container walls are perceived macroscopically as pressure. The temperature of the gas measures the average energy of motion (*kinetic energy*) of these particles. The higher the temperature, the more kinetic energy these particles have on average, and the higher their average speed is. When, in conditions of higher density and lower temperature, these particles come closer to each other, they begin feeling some mutual attraction caused by electric charges in the molecules and they cannot run far away from each other before hitting another molecule of the same kind. Then we have a liquid instead of a gas. The particles of the liquid can still move around each other

Fig. 3.8   Morning fog (Catinaccio range, Dolomites, Italy).

but do not travel far. At a fixed pressure, the phase transition from gas to liquid happens suddenly when the temperature is lowered to a critical temperature called the *condensation point*. If the temperature is lowered further and further, the particles have progressively less and less energy of motion and feel each other's electrical attraction more and more, until solidification occurs when the temperature drops to another critical value, called the melting point.[6] In the solid phase, the molecules cannot run away from each other but instead they form a structure in which their motion is restricted to oscillations around positions of equilibrium. The amplitude of these oscillations is larger if the temperature is higher (but below the melting point) and decreases as the temperature is lowered. These oscillations never stop completely though: the complete absence of motion

---

[6]An exception is liquid helium. Since in the atoms of this monoatomic noble gas the electric charges are not displaced and are distributed with spherical symmetry, there is no electric dipole moment and no forces between atoms arise to make bonds, so liquid helium flows freely without internal friction and does not solidify.

is forbidden by quantum mechanics [Griffiths (2005); Gasiorowicz (2003); Schiff (1968); Messiah (1961)], which rules the microscopic world of atoms and molecules. Physicists model the oscillations of a particle around a position of equilibrium with a harmonic oscillator which, in the macroscopic world, would just be a mass attached to a spring or an elastic. Only, because molecules are so small, their vibrations need to be described with an oscillator in quantum mechanics [Griffiths (2005); Gasiorowicz (2003); Schiff (1968); Messiah (1961)]. In this model, it was discovered that this oscillator has only a discrete set of possible values for its energy and that there is a lowest energy level which is necessarily non-zero. In other words, this oscillator never stops because it has always some non-zero energy of motion. Atoms and molecules cannot stop vibrating.

Perhaps not so obvious, there are also the direct change of phase from solid to gas (*sublimation*) and the inverse process. Particles can leave the surface of a solid and travel directly into the gas phase without going through the liquid phase. *Vice-versa*, particles in the gas phase can stick to the surface of a solid and remain there, like the molecules of water vapour sticking to cold rocks and covering them with verglas. Phase changes are very important for the mountain climber when it comes to water. Above a certain elevation, precipitation is in the form of snow instead of rain, and snow is made of ice crystals. The phase transitions of these water molecules become very important inside a snowpack because they are one of the main factors determining its stability and the risk of an avalanche. A snow slope can be safe early in the morning when the temperatures are lower and avalanche in the early afternoon. There are classic climbs where this phenomenon is known to happen regularly and one must descend before the afternoon avalanche hazard.

Another manifestation of phase changes in snow is the melting and weakening of bonds in snow bridges covering crevasses (Fig. 3.3). On the way to a high peak on a glacier, at night, some bridges support the weight of a climber and can be crossed relatively safely while, returning to camp in the afternoon heat, they collapse initiating what is, at best, an entertaining self-rescue or, at worst, a tragedy. Infrared radiation from the sun and the rising temperatures can melt snow efficiently: in spring and summer it is easier to cross creeks in the morning, while it can be much harder in the late afternoon when they are swollen with meltwater from higher up.

What about sublimation? In dry conditions, the passage of water molecules directly from the ice to the vapour phase is responsible for much of the disappearance of the snowpack in spring, together with melting. In polar regions, where the temperature is below 0°C, there is no melting but there is some water vapour in the air although, admittedly, it is so dry and there is so little precipitation that polar regions are rightly classified as deserts by geographers. It is sublimation that is responsible for the presence of this little moisture in the air.

The inverse phase change, from gas to solid, is more obvious in the phenomenon of frost, when water molecules from the air stick to cold surfaces to form ice.

In order to cause a phase change in a substance (and we refer here to chemically pure substances) one must supply or remove heat energy. Obviously, the amount of heat Q necessary is proportional to the mass m of the material which undergoes a phase change,

$$Q = Lm,$$

where the proportionality constant L is called a *latent heat of transformation* and has the meaning of heat energy that must be supplied or removed to make one kilogram of that substance change phase. For example, the latent heat of fusion of water ice is approximately $L_f = 334 \cdot 10^3$ J/kg and the heat necessary to melt 100 kg of ice at the standard pressure of one atmosphere is

$$Q = L_f m = \left(334 \cdot 10^3 \text{ J/kg}\right) \left(100 \text{ kg}\right) = 3.34 \cdot 10^7 \text{ J/kg}.$$

For a chemically pure substance, the temperature remains constant during the phase transition. The latent heat is given at the standard pressure of one atmosphere, as is the critical temperature at which the phase transition occurs. The critical temperature is different at different pressures. A familiar example comes up when we squeeze powder snow hard in our hands to make a snowball. The snow melts and makes a snowball even if the temperature of this cold snow is below 0°C. Similarly, at the bottom of a glacier, the large pressure of the ice on the lower layers sitting on bedrock can melt them even though the temperature is around −40°C, causing liquid water to lubricate the rock-ice interface. It is well known that glaciers flow under their own weight, with speeds ranging from centimeters to a meter per day, and melting due to pressure can be a factor in this process. A simple lab experiment further demonstrates this phenomenon: suspend a block of ice by its ends keeping it at a temperature below 0°C and lay a string through its middle section, then hang two equal weights (of mass m each), one to each end of the string (Fig. 3.9).

As time goes by, the pressure of the string melts the ice block and cuts through it. This pressure is the total weight $2mg$ divided by the area of contact, which is approximated as the length l of the string touching the top of the ice block times the diameter $2r$ of the string:

$$P = \frac{mg}{lr}.$$

Since the cutting process is slow, the melted water refreezes and the string appears to pass through the block while it is actually cutting it in two, until it reaches the bottom of the block and drops out of

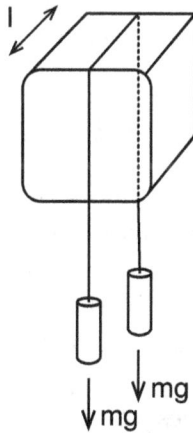

Fig. 3.9 The ice cube experiment.

it. The high pressure $P = \frac{W}{lr}$ is responsible for melting the ice at temperatures below 0°C.

The latent heat of melting, or evaporation, or sublimation is energy supplied to the molecules and is stored into the mass. It is given back during the opposite phase transition. For example, solar radiation evaporates ocean water, mostly at the tropics. This energy is stored in the moisture and, by transporting warm humid air, winds are responsible for a large chunk of the energy transfer that occurs in the atmosphere of the planet. This latent energy stored in the water vapour is released when it condenses, and it is the energy source which fuels hurricanes. Hurricanes cannot sustain themselves for long on land but, as long as they stay on water, they can draw more energy from water vapour. Similarly, evaporation cools down a surface because the evaporating particles take heat energy from the surroundings, which is used to break the bonds keeping these particles together in the liquid state. If we emerge on a windy ridge sweating from the ascent, the wind cools us down very quickly and we naturally cover up. If one fills up a white gas stove at −30°C one must be careful not to spill gas on one's hands because, at these temperatures, the cooling due to evaporation can easily give frostbite (the same can happen filling the car tank at the pump station on winter days in cold climates). That's a liquid to solid phase change of body moisture, a perennial risk of high altitude mountaineering. Phase changes are important in the mountains.

3.6   Phase changes at work: falling rocks

It has happened to everybody climbing icy peaks: rocks falling toward us from ridges or faces, sometimes spoiling an otherwise perfect sunrise or sunset. On high peaks there is plenty of loose rock in precarious equilibrium on ridges and faces that are severely weathered, and it takes very little to dislodge them. Water has a few peculiarities among common substances on the planet and one of these is that water *expands* when it freezes, contrary to most common substances which contract instead when going from the liquid to the solid phase. A mass of water turning into ice occupies more volume than it did in the liquid form. Its density (mass per unit volume) decreases by about 10%. An iceberg floats on water and what shows is just the "tip of the iceberg", as we see sometimes in alpine lakes or as it occur on a much bigger scale in the Arctic and in Antarctica. Water drips, runs, and accumulates in the small crevices between those loose rocks and the surrounding rock surfaces, and freezes there when the temperature drops. This process may have the effect of pushing the rock away from a precarious position of equilibrium, which happens when the temperature drops below zero degrees Celsius, for example at sunset. The forces arising in these situations when a solid is compressed are usually very large and can split boulders (Fig. 3.10).

Depending on the geometry of the situation, the ice thus formed may actually stabilize a wet rock and help holding it in place but then, when the first rays from the rising sun melt the ice that locks the rocks in place, the bond disappears and the rock may fall. Probabilistic considerations may help here. Even though only a small fraction of the rocks on a certain ridge is in precarious equilibrium,

Fig. 3.10   Weathering of granite (White Mountains, New Hampshire).

and only a small fraction of them may be held in position by ice, there are usually enough of them that some *will* fall. Think of climbs on those rotten ridges that nobody understands how they stay there in the first place — well they don't really stay there, they fall down a bit at a time. Even if only 0.01% of the loose rocks fall at the phase change, this means one hundred out of a million loose rocks will fall and a single one on the right trajectory is enough to hurt.

In general, water seeping into the cracks of rocks, roads, bridges, or concrete walls also causes substantial damage when it freezes and is a significant erosion agent in the mountains when the temperature oscillates around 0°C. One must be especially alert for falling rocks from snowy or icy places at sunrise and sunset and at any time when the temperature oscillates around the zero Celsius.

To have an idea of the forces at play when water freezes, let us estimate the pressure applied to the walls of a container filled with water when this turns into ice. The percent volume change of ice under compression is

$$\frac{\delta V}{V_0} = \frac{1}{B}\frac{F}{A},$$

where F is the magnitude of the force on the normal area A of the ice and $B = 1.13 \cdot 10^6 \, \text{N} \cdot \text{m}^{-2}$ is the bulk modulus of ice. When the mass m of liquid water freezes, the densities before and after the phase change must satisfy the relation

$$m = \rho_{\text{w}} V_0 = \rho_{\text{ice}} (V_0 + \delta V),$$

from which one obtains

$$\frac{\rho_w}{\rho_{ice}} - 1 = \frac{\delta V}{V_0}.$$

Knowing that the densities of freshwater and ice are

$$\rho_{\text{w}} = 1.00 \cdot 10^3 \, \text{kg} \cdot \text{m}^{-3},$$

$$\rho_{\text{ice}} = 9.17 \cdot 10^2 \, \text{kg} \cdot \text{m}^{-3},$$

respectively, this equation gives the pressure

$$P = \frac{F}{A} = B \frac{\delta V}{V_0} = B \left(\frac{\rho_w}{\rho_{ice}} - 1 \right)$$

$$= \left(1.13 \cdot 10^6 \, \text{N} \cdot \text{m}^{-2}\right) \cdot \left(\frac{1.00 \cdot 10^3 \, \text{kg} \cdot \text{m}^{-3}}{0.917 \cdot 10^3 \, \text{kg} \cdot \text{m}^{-3}} - 1 \right)$$

$$= 1.02 \cdot 10^5 \, \text{N/m}^2.$$

This pressure can certainly send a loose rock hurling down a wall.

3.7 Of steam and water bottles

Phase changes involve water bottles, too. Every mountaineer had liquid water turning into solid ice in the water bottle forgotten outside of the sleeping bag at night. But water turning into steam is worse, as my friends Carol and John swear.

After a day of strenuous backcountry skiing, Carol and John entered a cabin in the mountains of New Hampshire. While they settled in, John quickly started the woodstove and Carol pulled items out of her backpack. Tired, she had a long drink from her water bottle and, without thinking, she placed it on the stove which was still cold. They started the gas cooker, had dinner, and soon afterwards, they went to sleep very tired. The next morning, still half asleep, Carol opened her water bottle and a jet of steam hit her on the face.

When water turns into steam, one liter of liquid changes into 1600 liters of vapour at atmospheric pressure. The weak bonds keeping the molecules of liquid water close to each other, and due to attractive electric forces between unlike charges in these polar molecules, are broken during the phase transition. As liquid water turns into steam, all the energy supplied is spent breaking these bonds and the kinetic energy of the molecules is not increased. This is the reason why the temperature, which measures the average kinetic energy of the particles, does not increase during the phase transition. The steam temperature can begin to rise only after all the water has turned into steam. This amazing change of volume is what makes the steam engine so efficient and it is one of the basic reasons why the steam engine propelled the Industrial Revolution in the 1800s, changing Western civilization. Carol's experienced directly on her face this peculiar physical property of water that she had forgotten about. They skied out and eventually visited the nearest emergency ward. Luckily, the burn was very small and localized and there was no serious consequence. Phase transitions and water bottles don't go very well together.

3.8   Snowflakes and snowpack: the whole is more than the sum of its parts

Snow is made of ice crystals which combine to form snowflakes. These delicate geometries come in many varieties, usually based on a regular hexagon, and even a simplified analysis of their geometries would fill an entire book (in fact I know of one [Libbrecht and Rasmussen (2003)]). Ice crystals in a snowflake exhibit a six-fold symmetry, which is a direct consequence of the geometry of the water molecule (Fig. 3.1). Moreover, the geometry of a snowflake keeps repeating itself at smaller and smaller scales and such fractal geometries are of great interest to mathematicians [Peterson (1990)]. The geometry and physics of snowflakes are essential to understand the formation and evolution of a snowpack. Mountaineers and skiers are interested in the stability of the snowpack: nobody wants to trigger an avalanche. Snow science is very complex and takes years to learn (see [Fredston and Fesler (1988); Daffern (1992)] for introductions).

Two processes occur after a snowfall: *settling* and *metamorphism*. Settling is the process by which new snow compresses old snow and becomes denser under its own weight. Settling always increases the stability of the snowpack. Metamorphism, which comprises settling, sintering, and melt-freeze metamorphism, is more complex and depends on the temperature gradient within the snowpack. The magnitude of the temperature gradient ∇T is, approximately, the temperature difference ΔT between two points divided by their distance Δz,

$$\nabla T \simeq \frac{\Delta T}{\Delta z}.$$

More precisely, the temperature gradient $\vec{\nabla}T$ is a vector with magnitude equal to the limit of $\Delta T/\Delta z$ when the separation Δz between the two points becomes very small, $\Delta z \to 0$, and with direction pointing in the direction of maximum temperature increase. This property reflects a general fact in mathematics. Given a function $f(x, y, z)$ of the position, the gradient

$$\vec{\nabla}f = \left(\frac{\partial f}{\partial x}, \frac{\partial f}{\partial y}, \frac{\partial f}{\partial z}\right)$$

is perpendicular to the surfaces of constant f (along which, by definition, there is no variation of f and its partial derivatives are zero) and points in the direction of maximum increase of f.[7]

We are interested in the temperature difference between the ground, which is usually near $0°C$, and the top layers or the surface of the snowpack, which can be quite colder (remember that snow acts as a thermal insulator for the ground underneath).

When the temperature gradient is moderate, the two major metamorphic processes are *rounding* and *sintering* of ice crystals. In rounding, the individual snow crystals lose their angular features and become more rounded, while in sintering these crystals bind together by forming necks at the contact points. Both processes make the snowpack more stable. These processes are due to sublimation: water molecules leave the convex ice surfaces and are deposited on concave ones.

When the temperature gradient within the snowpack is large, *faceting* takes place instead. Water molecules rise from the warmer bottom layers of the snowpack and re-crystallize on the cold top layers, transforming ice grains into faceted grains. If the conditions are right, very large crystals grow in the snowpack, forming the *depth hoar* familiar to mountaineers and skiers. These large depth hoar crystals are poorly bound to their neighbours and form a weak layer

---

[7]As an example, if $h(x, y)$ is the elevation as a function of the position on a map, the gradient $\vec{\nabla}h$ has no components along contour lines (*i.e.*, lines of constant h) and points perpendicular to them. The magnitude of $\vec{\nabla}h$ is largest where the variation of elevation is maximum, which happens near rock walls where contour lines on the map pack closely together.

in the snowpack. When the water molecules re-crystallize on the surface of the snowpack (or on any cold surface), they form *surface hoar*, which can consist of very large crystals up to a few centimeters in size. Surface hoar can form a great surface to ski on but, when it is buried by new snow, often it does not change for long periods of time, and possibly through the entire winter. This surface hoar is not bonded to the layers above and below and forms a weak layer, a sliding surface which can persist all winter. Given the right conditions, rounding and sintering can metamorphosize depth hoar or buried surface hoar. However, a single night of intense cold can be sufficient to form depth or surface hoar. The processes of rounding/sintering and of faceting compete with each other and stabilize or destabilize, respectively, the snowpack.

When, in spring and summer, the temperature at the surface of the snowpack is sufficiently high to melt the snow during the day but night time temperatures fall below freezing, *melt-freeze metamorphism* begins. Water percolating from the top layers refreezes and the result is the formation of large ice crystals forming the corn snow of spring skiing. When the temperature reaches $0°C$ the bonds between ice crystals deteriorate and the snowpack becomes unstable. As the temperature drops, the snow becomes more stable as new bonds are formed between crystals. The snowpack may be stable early in the morning and become unstable in the middle of the day. A macroscopic experiment on snow crystals involuntarily performed by many climbers, which emphasizes the melt-freeze metamorphism, is crossing a snow bridge on a crevasse. Early in the morning the bridge holds the weight of a climber, while later in the day it fails with consequences ranging from unpleasant to dramatic.

Does it sound complicated? If it does, it's probably because it is. And all this is for a snowpack left alone, or interacting with its environment in a minimal way only through temperature gradient, humidity, and the occasional snowfall. What if it is disturbed by rain or wind? Rain on a snowpack generally weakens it. However, if it is followed by freezing, it may make it stronger instead. Rain early in the winter, which freezes on the surface of the snowpack, may glaze it with the consequence that later snow accumulating on it may not bond well and this sliding surface may continue to exist all winter. Wind can compact the top layers of snow, which can be cold and bind poorly to the layers below, forming a wind slab.

The snowpack may change daily and all the events and changes which occurred since the first snowfall may be important. Skiers and mountaineers dig snowpits to assess the conditions of the snowpack, identify layers, measure their strength and the strength of the bonds between layers (see, *e.g.*, Fredston and Fesler (1988); Daffern (1992), but these references are no substitutes for a proper avalanche safety course). However, no single test is sufficient to draw conclusions and even snow scientists make occasional mistakes in the interpretation of results from their pits. Isolated snowflakes are beautiful complicated geometries which require work to understand, while a snowpack which is composed of a large number of these snowflakes is a complicated dynamical system which interacts with its environment in a variety of ways. The whole is more than the sum of its parts.

3.9   Avalanche!

In the old days in the European Alps avalanche, the "white death", was the nightmare of locals and travellers. Avalanches threaten road and railway traffic and sometimes villages, and they shut down mountain roads for the entire winter. For the ski-mountaineer, the mountaineer, and for people living in the mountains it is a recurrent danger which requires knowledge, skills, and practice to detect and to avoid. In more recent history, in North America avalanches have taken their toll on snowmobilers who travel faster and have less time to detect danger, in addition to the fact that being on a motorized vehicle rather than relying on one own's ambulation makes one less prone to detecting nature's warnings. Avalanches happen naturally, but most avalanche-related accidents involve human-triggered avalanches and the understanding of snow and avalanche plays a major role in avoiding such accidents. At least three factors concur to create avalanche conditions; terrain, snow, and weather. Judging avalanche hazard is a science and an art which requires study and dedication, and anybody living or going in the mountains in winter should develop a sense for snow stability and obtain instruction. Here we comment only on a few physical aspects of the physics of avalanches. See Fredston and Fesler (1988); Daffern (1992) for more information.

There are various kinds of avalanches: loose powder avalanches, wet snow avalanches, slab avalanches, and possibly something in between but, to be schematic, these are the three types usually identified. One aspect that may be very important is how fast avalanches travel: their speed is very different according to their type. A wet snow avalanche from late spring or summer is more like a pile of blocks that roll and slide down slowly. However, they are so heavy that they crush things in their path. A fresh powder avalanche, like those seen in winter around Davos in Switzerland or Rogers Pass in Western Canada, can travel much faster because the snow has much air in it and the physics of its motion is quite different. Finally, a slab avalanche has a rather curious dynamics: a layer of consolidated snow sitting on top of a weaker layer fractures, the fracture travels across the slope, and the plates which formed the top layer slide down the slope, breaking up in a maze of smaller blocks. A typical situation is that of a strong layer of snow compacted by wind which sits on top of a layer of deep hoar crystals ("sugar snow") formed earlier on, which got buried by a new snowfall but never bonded to the top layer.

The snowpack may also be sufficiently stable that avalanches do not occur on their own at a particular moment of time, but when stress is applied by a skier, mountaineer, snowmobile, or chamois, the snowpack fails.

The risk of avalanches is intimately related to the snow cycle. The strength of the bonds between ice crystals is essential to determine whether a layer of snow is weak or strong. Then the bonds between layers become extremely important to assess the stability of the snowpack. Two distinct layers which are weakly bonded to each other create avalanche conditions. A strong layer sitting on top of a weak one creates the potential for a slab avalanche. The conditions of the snowpack vary in time during the season, and from season to season, and one must be aware of this dynamics. Spatially, the structure of the snowpack can also change within short distances, and can change dramatically according to the orientation of the slope and to the direction of the dominant winds which transport snow.

The amount of snowfall recorded during a period of bad weather, or during the winter, is only an indication of the actual thickness of the snowpack encountered on a trip, because wind also transports snow and can be very efficient at loading slopes. Characteristic "pillows" form on the lee side of a windy mountain which is weighted, while the slopes on the opposite side are unloaded by the wind. It helps to know the dominant winds and what kind of snow fell previously: cold, light powder which is easily picked up and transported by winds, or heavy wet snow which sticks where it falls.

Settling of the snow in the snowpack is largely affected by its temperature. Generally speaking, the warmer the snow, the faster the settling, which involves breaking down the crystals which lose their star shapes and become more and more rounded, bonding with neighbouring crystals. By contrast, this process is slower in a colder snowpack. Another phenomenon is the development of weak layers of temperature gradient snow: if there are substantial differences in the temperatures of two layers of snow (a temperature gradient), weak layers known as *depth hoar* or sugar snow can develop in low-density snow. This layer may remain buried and it does not bind to the neighbouring layers for the entire winter and forms a sliding surface for the top layers. Similarly, a period of cold dry weather can form large hoarfrost crystals on the surface of the snow (surface hoar). Once this layer is buried by new snow, it will create avalanche conditions.

Signs of unsettled snow include large lumps of snow sitting on tree branches. As the settling progresses, snow progressively falls off the trees. A snow pit analysis, if dug sufficiently deep, reveals the weak layers, but it must have the same exposure and slope as the main slope that one is trying to assess. Skiers and mountaineers do develop knowledge of avalanches, but they try hard to stay away from them. The avalanche patrol members, instead, look for avalanche terrain and avalanche conditions all the time and they have the job of taking risks and triggering avalanches to make ski slopes and roads safe. For practical purposes, they are the most knowledgeable persons about avalanche. Snow scientists have both theoretical and experimental knowledge of the snow cycle and of avalanche dynamics.

In areas where avalanches cannot be avoided, it is sometimes possible to protect railways and roads with tunnels, or setting up fences intended to stop or reduce the avalanches. Some prevention may help, but it is of limited use. Paul Preuss, who was not only an exceptional free climber but also an expert of mountains in their various aspects, remarks in his writings how long grass on steep slopes not cut by villagers and flattened by rain facilitates the sliding of snow, which does not happen when the grass is short and the snow binds better to the soil [Preuss (1986)].

3.10   Suncups and penitentes

Walking up a snow slope in spring or summer we often notice fields of bowls on the surface of older, settled snow, of size ranging from a few centimeters to half a meter, which are very handy to step on and can be used as natural steps on the hard surface, especially useful if the slope is a bit steeper: *suncups*. These suncups are wider than they are deep and the physical mechanisms by which they form and grow are quite interesting. It is believed that their origin is the same as the much more dramatic structures called *penitentes* encountered at high elevation in the Andes or the Himalayas, blade-like structures of snow or ice wider at the base and narrower at the tip, which can be up to six meters tall. Fields of penitentes have been likened to a procession of penitent monks in white robes, hence their Spanish name. The first report of penitentes in the literature seems to be the one by Charles Darwin himself in his *Voyage of the Beagle* [Darwin (1989)], where he describes passing through penitente fields in the mountains of Chile.

The origin of these structures is believed to be the process of *ablation*, the removal of snow by sublimation, the passage of water molecules from the solid to the vapour state without going through the liquid phase, which we have already described in Sec. 3.5. Sublimation occurs due to the intense sunlight in dry, cold conditions. Ultraviolet rays do not play a role: most of the absorption of solar radiation by snow happens in the infrared. Once the formation of the peaks has begun, a hollow receives much more radiation than the nearby peaks due to reflection and an instability sets in by which the through becomes deeper and the structure grows taller and taller. Remarkably, the same kind of penitente structure is observed in physics laboratories on surfaces irradiated with laser light, where microscopic penitentes 10 to 100 microns in size (10^{-5}–10^{-4} m) are formed. Important factors are the presence of dirt on the surface of the snow and its thickness. A thin layer of dirt inhibits reflection from the snow but transmits heat and impedes the formation of structure by reducing temperature gradients, while a thick layer of dirt stops reflection and preserves the snow underneath, favouring the formation of penitentes. The detailed theoretical modelling of the process is fairly complicated [Betterton (2001)] and the process is also studied in the laboratory, where small penitentes with a size of a few centimeters have been grown [Bergeron, Berger, and Betterton (2006)].

A related structure is the *dirt cone*, which is formed when dirt protects the top of a cone of snow or ice from ablation, while the sides get eroded. The dirt becomes more concentrated on the top and the process accelerates until a balance is reached between growth and grains of dirt sliding off the top. These structures can be quite high, the tallest reported being an 85 m cone observed in the Himalayas [Swithinbank (1950)].

Chapter 4

Glacier puzzles

4.1 Introduction

Glaciers are beautiful physical systems created by the accumulation of snow and its subsequent compression, packing, and metamorphosis into ice. This ice flows, cracks at crevasses, erodes its bed and the supporting valley walls, and pushes moraines at its front and sides (Figs. 3.5, 4.1, 5.2). There are many interesting phenomena taking place in glaciers which show up visually, sometimes spectacularly, in their morphology [Post and Lachapelle (2000)]. There is much physics involved in the experimental study of glaciers and polar ice sheets and ice caps, and in their theoretical modelling. In order to understand glaciers, one should first of all study the microscopic and crystalline structure of ice, the behaviour of ice as a material, and its response to stresses, and then the flow of ice as a slow fluid, ruled by the notoriously difficult Navier-Stokes equations of fluid mechanics. There is a mass balance for glaciers which arises from the accumulation of ice and its removal by melting (*ablation*). Next there is thermodynamics: the glacier receives heat from radiation and other sources, some ice melts, and energy is exchanged. Then there is lubrication at the bottom. Further, glaciers have their own hydrology, which is complicated by the fact that meltwater and rainwater can freeze. There can be ice quakes and seismic waves in the ice.[1] Then, statistical analysis

---

[1]See [Ekstrom, Nettles, and Tsai (2006); Tsai and Ekstrom (2007); Tsai, Rice, and Fahnestock (2008); Failletaz, Funk, and Sornette (2009); Górski (2014)].

is done on data from glaciers and ice sheets, and on and on. Glaciology is a science in itself, it has many facets, and its own textbooks (*e.g.*, [Paterson (1994); Cuffey and Paterson (2010); Hooke (2005); Bennett and Glasser (2009); van der Veen (2013); Greve and Blatter (2009)]) but it is not a finished science and much remains to be discovered. Let us address some questions that arise naturally when wandering in glaciated mountains.

4.2 How thick can an alpine glacier be?

How thick can an alpine glacier be? Measurements show that alpine, or "valley", glaciers are tens to hundreds of meters thick. The ice doesn't become thicker than this value, so some basic physics sets this limit to the glacier's thickness. By contrast, polar ice sheets can be a few kilometers thick, the highest thickness measured being nearly 4.8 km in East Antarctica [Paterson (1994)].

For an alpine glacier which sits on a slope, the maximum thickness of the ice is limited by the tensile stress of this material. A glacier deforms and flows, but not without limit. There are two forces at play: the component of the weight of the ice along the slope and the internal stress in this material, which keeps the ice together (Fig. 4.1). The ice flows, within certain limits, under the pull of gravity but when the component of its weight directed along the slope exceeds a certain limit (the tensile strength), the ice breaks off. This description is simplistic because melting at the base of the glacier lubricates the motion of the ice relative to its bed and promotes sliding, but for simplicity we will ignore this fact here.

Is it possible for an alpine glacier on a slope to grow to the point of breaking? In the European Alps there are historical records of glaciers breaking off. The Altels breakoff of 1895 in the Valais region of Switzerland was truly colossal [Failletaz, Sornette, and Funk (2011)]. It is estimated that $4 \cdot 10^6 \, \mathrm{m}^3$ of ice fell, causing what is still the largest known ice avalanche in the European Alps. The Allalin glacier, again in the Valais, broke off repeatedly in the past two centuries. Tragically on August 30, 1965, it broke again and caused

Fig. 4.1 Internal stresses keep these ice blocks attached (Mont Blanc, France).

88 victims among the crew working on the construction of the Mattmark hydroelectric plant dam 400 m below. It is estimated that $2 \cdot 10^6 \, \mathrm{m}^3$ of ice fell [Failletaz, Funk, and Sornette (2012)]. Although simplistic, it is not unrealistic to consider that a glacier on a slope can break under its own weight and fall.

Deriving an upper limit for the ice thickness of a valley glacier is basic material from glaciology textbooks. Approximate the glacier with a parallel-sided block of ice on an incline making an angle α with the horizontal (Fig. 4.2).

Let h be the thickness of the block and A be its basal area. The component of its weight pointing along the slope tends to shear this block and, if the latter becomes too thick, the maximum tensile stress which keeps the block attached to the rest of the ice will not be able to balance gravity so the block will detach and fall. Assume that the forward motion of the glacier is due entirely to the internal deformation of the ice (that is, neglect basal sliding).

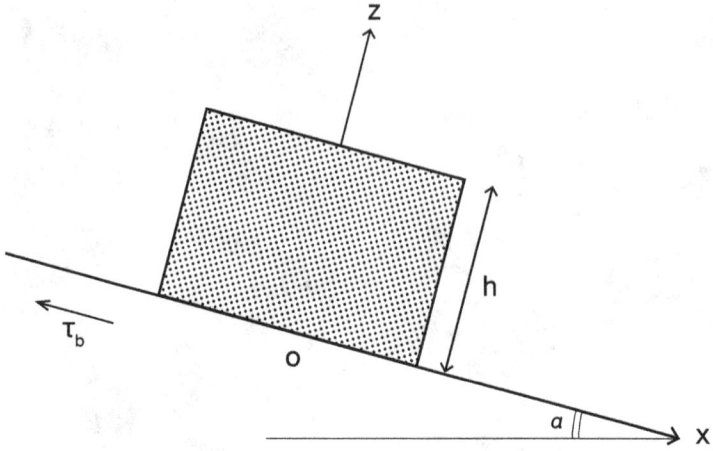

Fig. 4.2   A block of ice of thickness h on an incline of slope angle α.

As we have already seen in Sec. 3.4, shear stress is a force divided by area, very much like pressure, but with an essential difference. While pressure is a force per unit of area *perpendicular* to the force, shear stress is a force per unit of area *parallel* to the force (think of pushing a table with a force parallel to the table top and shearing it, as in Fig. 3.6).

The shear stress multiplied by the basal area A gives the force holding the block in place. It points parallel to the incline but upwards, opposing gravity (Fig. 4.2). The intensity of the force from the shear stress is $\tau_b A$, where τ_b is the value of the ice basal stress. The weight of the block has a component along the incline $mg\sin\alpha$, where $m = \rho A h$ is the mass of the block and ρ is the density of the ice. This mass is the density times the volume Ah of the ice block. When these two forces just balance, the ice is as thick as it can get:

$$A\tau_b = \rho g h A \sin\alpha,$$

which gives the maximum thickness

$$h = \frac{\tau_b}{\rho g \sin\alpha}. \tag{4.1}$$

Since a small slope angle α means a small value[2] of $\sin \alpha$ and $\sin \alpha$ appears in the denominator, it follows from this simple model that the ice is thin where it is steep, and can be thicker on milder slopes, which is rather obvious to mountaineers.

What about the numerical value of h? A typical value of the basal stress in temperate[3] alpine glaciers is $\tau_b \simeq 10^5$ Pa [Paterson (1994); Cuffey and Paterson (2010)], which yields

$$h = \frac{11 \text{ m}}{\sin \alpha}.$$

The 11 m figure may seem small, but remember that $\sin \alpha$ is always smaller than 1 (unless α is ninety degrees, which means a vertical wall instead of a slope) so $1/\sin \alpha > 1$ and it grows quickly for small slope angles α. This estimate is corrected for valley glaciers, for which part of the weight of the ice is supported by the valley walls. This is done "by hand" by introducing a so-called *shape factor* (a pure number without units) $f \simeq 0.5 - 0.9$ in Eq. (4.1) [Paterson (1994); Cuffey and Paterson (2010)], which becomes

$$h = \frac{\tau_b}{f \rho g \sin \alpha}. \tag{4.2}$$

When the slope angle is small, so that $\sin \alpha \approx \alpha$, one can approximate the formula as

$$h \simeq \frac{\tau_b}{f \rho g \alpha} \simeq \frac{11 \text{ m}}{f \alpha}.$$

Glaciologists are more interested in obtaining a numerical value for the basal shear stress τ_b from the known ice density and acceleration of gravity and the measured slope α because it is technically difficult, or practically impossible, to access the base of the glacier to measure τ_b directly. However, this can be done in rare situations when there are ice caves or artificial tunnels at the base of the glacier.

---

[2]The sine function is linear for small values of its argument.
[3]Technically, a *temperate glacier* is defined as one which is at the melting point throughout the year, from its surface to its base. This is in contrast with a *polar glacier*, which is always below melting point from surface to base.

4.3   How thick can a polar glacier be?

A giant ice sheet or polar ice cap is much thicker than an alpine glacier (kilometers versus tens or hundreds of meters) and buries the underlying topography, so that in practice it is considered to be horizontal by glaciologists. The thickness of the ice is much larger than the elevation difference between top and bottom of the slope on which the glacier sits. In other words, the thickness of a polar ice cap is not determined by the underlying topography, as for alpine glaciers, which are dominated by it.

Consider the zero slope limit $\alpha \to 0$ of Eq. (4.2) of the previous section which expresses the maximum thickness of an alpine glacier on a slope. In this limit, since $\sin \alpha \to 0$ as $\alpha \to 0$, the maximum ice thickness h becomes infinite, which means that gravity does not set limits on this quantity on a flat surface. However, the thickest ice sheet measured on Earth is less than 5 km thick, so some other physics sets an upper bound on h. We have already seen in Sec. 1.2 Weisskopf's physical argument providing the maximum height of a mountain: if material keeps being added to its top until it exceeds the maximum possible height for that material, the mountain begins to melt at its base and to flow. Now the same reasoning can to be applied to glacier ice instead of rock, which means using the numerical value of the latent heat of fusion of ice instead of that of rock.[4] The specific heat of fusion of ice is $L_f = 3.34 \cdot 10^5$ J/kg and, using again $\eta \sim 1/3$, our equation (4.1) yields the maximum thickness $h \simeq 11$ km [Faraoni and Vokey (2015)]. This is indeed the correct order of magnitude for polar ice sheets. It is only twice the measured value, which is good enough for an order of magnitude estimate — by definition an order of magnitude estimate is not precise.

4.4   How deep can a crevasse be?

Crevasses are both fascinating, because of the physics they are associated with [Colgan *et al.* (2016)], and intimidating because of their

---

[4]Indeed, many geology textbooks express the view that ice can be seen as a rock close to its melting point.

danger. Falling in a crevasse while crossing a glacier is a scary and dangerous experience and we always rope up on a glacier. How deep could a crevasse be? Crevasses are not arbitrarily deep and some physics must set an upper limit to their scary depth. Crevasses are fractures in the ice due to stresses originating as a glacier advances (Figs. 4.3, 4.4, and 3.3), and their theoretical modelling constitutes an application of the broad physical theory of fracture mechanics in solids [Lawn (1993)]. However, a much simpler physical reasoning provides easily an order of magnitude estimate for the maximum depth of a crevasse [Nye (1952)].

A crevasse forms in glacier ice when the tensile stress pointing perpendicular to the crevasse walls exceeds the tensile strength of ice. Suppose that a crevasse has opened down to a depth h. The simplest situation is the one in which two equal stresses of intensity σ contribute to widening, one on each crevasse wall, adding to a total stress 2σ (Fig. 4.5). The pressure of the ice (*cryostatic pressure*, analogous to the hydrostatic pressure in water due to the column of liquid above) tends instead to close the crevasse.

Fig. 4.3 A crevasse near the summit of a glacier, called *bergschrund*.

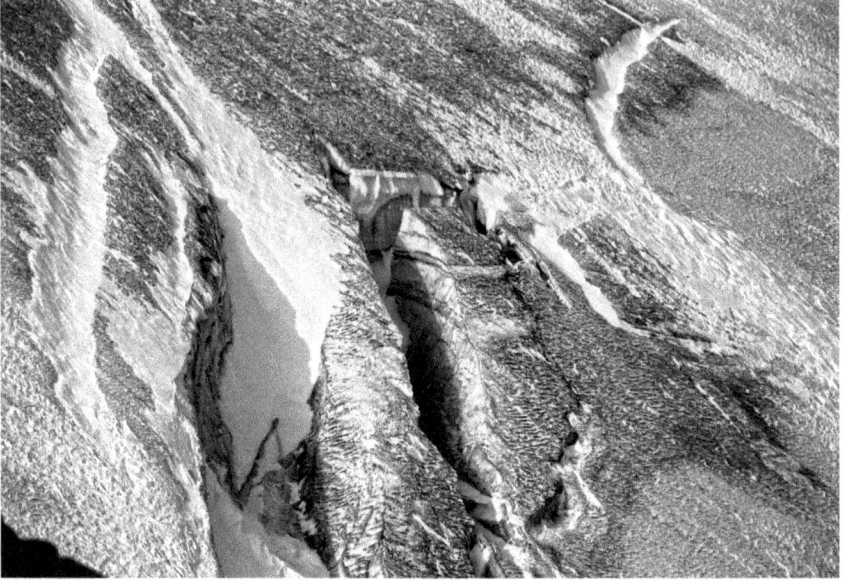

Fig. 4.4   Crevasses on a small alpine glacier.

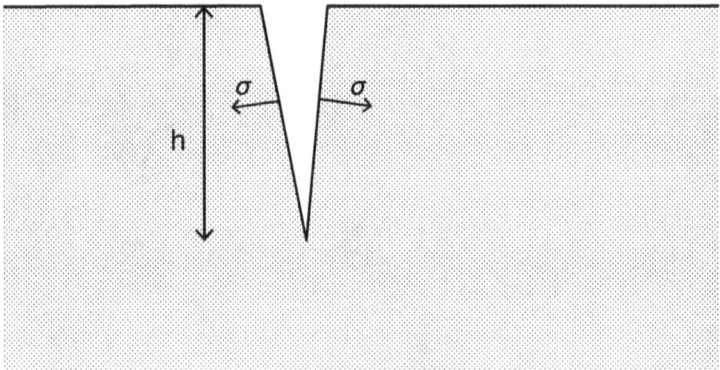

Fig. 4.5   Stresses acting on the walls of a crevasse.

The value of this pressure is $\rho g h$, where ρ is the ice density, g is the acceleration of gravity, and h is the crevasse depth. In this simple model, the maximum crevasse depth is reached when this cryostatic pressure balances the total sideways stress. In formulas, we have

$$\rho_i\, gh \simeq 2\sigma,$$

which gives

$$h \simeq \frac{2\sigma}{\rho_i g}.$$

Using the typical values $\sigma \sim 10^5$ Pa for the yield stress of ice, $\rho_i = 917\,\text{kg/m}^3$ for its density, and $g = 9.8\,\text{m/s}^2$ for the acceleration of gravity, one obtains $h \simeq 22\,\text{m}$. Crevasse depths in temperate glaciers are rarely reported in excess of 30 m, hence the simple calculation above is correct as an order of magnitude estimate. One should, however, take note of the fact that the yield stress increases as the temperature decreases, leading to the prediction that crevasses in polar glaciers, which are significantly colder than temperate glaciers, will be deeper. Larger crevasse depths are indeed reported in cold polar regions [Paterson (1994); Cuffey and Paterson (2010); Hooke (2005)].

4.5 How long does it take for snow to turn into ice?

The snow falling onto a glacier is covered by new layers of snow, pressed down, and compacted. We can see that, eventually, fluffy snow turns into hard ice. After this stage is reached, glaciologists usually model glacier ice as incompressible for most purposes. Two questions arise naturally. How long does this metamorphosis process take? What happens at the microscopic level?

Changes in the density of ice sheets can be assessed with modern interferometric radar techniques and neutron probes, which allow scientists to reconstruct the density profile in the top layers of the ice [Arthern *et al.* (2013); Hawley, Morris and McConnell (2008)].

At the microscopic level there are complicated physical processes involving rounding of the ice crystals, re-crystallization, sintering, grain settling, and packing by sliding [Benson (1962); Anderson and Benson (1963); Alley (1987)] which constitute an entire area of research in themselves. But can we say anything from the macroscopic point of view about the densification of[5] firn? To a certain extent, we can. Consider, for simplicity, a glacier on a horizontal bed, let z be the depth measured vertically from the ice surface, and denote with $\rho(z)$ the mass density at depth z. An approximation used in glaciology and known as *Sorge's law* assumes that, in conditions of constant temperature and constant snow accumulation, ρ is independent of time [Bader (1954)]. To simplify things further, one can assume that the snow is dry, although this is not always true. Liquid water due to melting and percolation and its refreezing are significant complications and it is better to ignore them in a first approximation. An empirical law providing the snow density as a function of depth is

$$\rho(z) = \rho_i - (\rho_i - \rho_s)\, e^{-cz}. \tag{4.3}$$

Here ρ_i is the density of ice and ρ_s is the density of snow at the surface, while the constant $c = z_0^{-1}$ gives us the spatial scale z_0 over which the firn density changes appreciably. This empirical law was reported long ago [Schytt (1958)]. The density of dry snow is approximately $\rho_s = 330\,\text{kg} \cdot \text{m}^{-3}$ [Paterson (1994)], which gives $\rho(z_0) = 0.7645\rho_i$, $\rho(2z_0) = 0.9134\rho_i$, and the value $0.99\rho_i$ of $\rho(z)$ is reached at $z = 4.16\,z_0$. In practice, we can assume that the density of ice is reached at depths $3\,z_0$ or $4\,z_0$. The exponential law (4.3) is plotted in Fig. 4.6. In this figure the density of snow is plotted in units of the ice density $\rho_i = 917\,\text{kg/m}^3$ and the depth z is given in units $z_0 = c^{-1}$. The surface density is $\rho_s = 330\,\text{kg/m}^3$, a value appropriate for dry snow areas.

Although the exponential law was derived empirically, for somebody who likes math it is fun to derive it from general considerations [Faraoni (2016)]. First, the density $\rho(z)$ is minimum at the surface $z = 0$, where it assumes the value ρ_S. Second, the density cannot be larger than the density of ice ρ_i. Third, $\rho(z)$ should always be increasing with the depth z, so its rate of change must have sign $d\rho/dz > 0$.

---

[5] *Firn* is partially consolidated snow that has experienced one or more summer melting seasons but is not yet glacier ice. Its density ranges between $400\,\text{kg/m}^3$ and $830\,\text{kg/m}^3$ [Paterson (1994); Cuffey and Paterson (2010)].

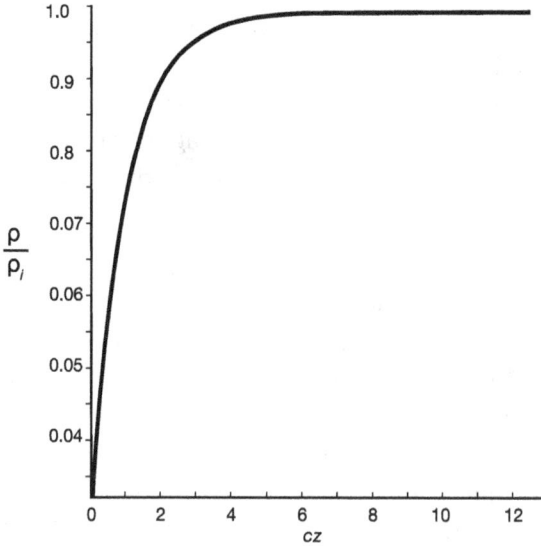

Fig. 4.6 The density of snow turning into ice (in units of the ice density) as a function of the depth z (in units of $z_0 = c^{-1}$).

$\rho(z)$ should approach its upper bound ρ_i either at a finite depth z_0, or perhaps as $z \to +\infty$ (as the mathematicians often allow and as the environmental scientists use in common diffusion problems [Boeker and van Grondelle (2011); Mason and Hughes (2001); Monteith and Unsworth (1990); Guyot (1998); Henry and Heinke (1989); Faraoni (2006)]) but in this case the approach to the limiting value must be fast. Formally, these ingredients imply that $\frac{d^2\rho}{dz^2} < 0$ and $\frac{d\rho}{dz}, \frac{d^2\rho}{dz^2} \to 0$ as $z \to +\infty$. The gradient of $\rho(z)$ is larger for loose snow near the glacier surface and decreases as the depth increases, until it goes to zero when the density is approximately that of ice and changes no longer occur, that is, the gradient $d\rho/dz$ is a function of the difference $\rho_i - \rho(z)$:

$$\frac{d\rho}{dz} = f\left(\rho_i - \rho(z)\right).$$

One then observes that the density does not change enormously when loose snow is transformed into ice. Dry snow at the surface of polar glaciers has density $\rho \sim 330 \, \text{kg} \cdot \text{m}^{-3}$, while for glacier ice $\rho \sim 830 - 917\,\mathrm{kg \cdot m^{-3}}$ [Paterson (1994)]. The range of densities involved is not large: it spans roughly a factor of three. This is not the case in many other geophysical processes in which the quantities involved change over several orders of magnitude. Then, we linearize the function $f(\rho_i - \rho)$, obtaining the approximate ordinary differential equation

$$\frac{d\rho}{dz} = c\,[\rho_i - \rho(z)], \tag{4.4}$$

where $c > 0$ is a constant with the dimensions of an inverse length. All the solutions of our problem must have $\rho < \rho_i$ and they automatically have $d\rho/dz > 0$.

If you know a bit about the theory of ordinary differential equations, now the mathematical solution of our problem is not difficult [Hille (1969); Brauer and Noel (1986)]. Following the standard solution method, the general solution of the homogeneous equation $d\rho/dz = -c\rho(z)$ is $\rho_{\mathrm{hom}}(z) = \rho_0\,e^{-cz}$, where ρ_0 is an integration constant. A particular solution of the inhomogeneous equation (4.4) is $\rho = \mathrm{const.} = \rho_i$. The *general* solution of the differential equation (4.4) describing our problem is the sum of these two solutions [Hille (1969); Brauer and Noel (1986)],

$$\rho(z) = \rho_0\,e^{-cz} + \rho_i.$$

The boundary condition $d\rho/dz \to 0^+$ as $z \to +\infty$ is already satisfied, as it is easy to verify. We then impose the other boundary condition $\rho(0) = \rho_s$, which fixes the integration constant to be $\rho_0 = \rho_s - \rho_i$ and gives the density profile as

$$\rho(z) = \rho_i - (\rho_i - \rho_s)\,e^{-cz}.$$

Now we can address the other question, of how long it takes for snow to turn into ice [Paterson (1994); Cuffey and Paterson (2010)]. Let $v = dz/dt$ be the vertical velocity of a small volume of firn with vertical position $z(t)$. Mass conservation for a fluid of density ρ and velocity v in a stream tube of cross-sectional area A gives the continuity equation (*e.g.*, [Halliday, Resnick, and Walker (2005)])

$$\rho v A = \mathrm{const.} = \rho_s v_s A_s.$$

Since there is no sideways motion, we consider a vertical parallelepiped of constant cross-sectional area $A = A_s$ as the stream tube and

$$\rho(z)v(z) = \text{const.} = \rho_s v_s,$$

where the constant quantity $\rho_s v_s$ is the accumulation rate of ice, which glaciologists express in convenient units of $\text{kg}/(\text{m}^2 \cdot \text{yr}))$. Since $dt = dz/v(z)$ and $v(z) = \rho_s v_s/\rho(z)$, the time taken for a volume of snow deposited at the surface to arrive to the depth z_i of the firn-ice transition is

$$
t_i \equiv t(z_i) = \int_s^{z_i} dt = \int_s^{z_i} \frac{\rho(z)}{\rho_s v_s} \, dz
$$

$$
= \frac{1}{\rho_s v_s} \left[\rho_i z + \frac{\rho_i - \rho_s}{c} \left(e^{-cz} - 1 \right) \right]_0^{z_i}
$$

$$
= \frac{\rho_i}{\rho_s v_s} \left[z_i + \frac{1}{c} \left(1 - \frac{\rho_s}{\rho_i} \right) \left(e^{-cz_i} - 1 \right) \right].
$$

The age of the ice at the depth z_i of the firn-ice transition can be measured empirically by counting annual layers of firn near the top of the glacier, much like annual growth rings in a tree. For $\rho_i = 917 \, \text{kg/m}^3$ and $\rho_s = 300 \, \text{kg/m}^3$ (for dry snow), the previous equation gives

$$
t_i = \frac{917 \, \text{kg/m}^3}{\rho_s v_s} \left[z_i + \frac{0.6728}{c} \left(e^{-cz_i} - 1 \right) \right]. \tag{4.5}
$$

This formula provides theoretical values for the age of the ice of glaciers, which can be compared with empirical measurements. It is instructive to look at the numbers for a couple of Arctic and Antarctic glaciers as an example.

Site A in Greenland has $\rho_s v_s = 265 \, \frac{\text{kg}}{\text{m}^2 \cdot \text{yr}}$, $z_i = 75 - 80 \, \text{m}$, $t_i = 185 \, \text{yr}$, $c = 2.35 \cdot 10^{-2} \, \text{m}^{-1}$ which gives, using Eq. (4.5), $t_i = 185 \, \text{yr}$, in good agreement with the measurements.

At Dome C in Antarctica, measurements have provided $\rho_s v_s = 36 \, \frac{\text{kg}}{\text{m}^2 \cdot \text{yr}}$, $z_i = 100 \, \text{m}$, $t_i = 1700 \, \text{yr}$, and $c = 1.65 \cdot 10^{-2} \, \text{m}^{-1}$. Our equation (4.5) gives $t_i = 1708 \, \text{yr}$, again in good agreement with the measured value. It takes 1700 years for the cold, dry polar snow to turn into ice! The process is much faster on alpine glaciers, where some liquid water is usually present.

4.6   How did glacial valleys get their U-shape?

Since when we were little kids we were told at school that glacial valleys are U-shaped (Fig. 4.7), while valleys carved by rivers are V-shaped instead (Fig. 1.1). This statement appears in most geography and geology books (*e.g.*, [Skinner and Porter (1992); Lutgens and Tarbuck (2000)]) and has been around since the inception of glaciology [Campbell (1865); McGee (1894)]. Mountaineers indeed confirm that glacier-carved valleys have characteristic U-shapes, and some will tell us also that the cross-sectional profiles that they carve in rock are usually steep.

The erosion of the ice on the valley walls and the glacier bed is truly remarkable, but it needs to last tens of thousand of years

Fig. 4.7   U-shaped glacial valley (Val Neigra, Trentino region of Italy).

to produce the valleys that we admire. How do glaciologists model
the transverse profile of a glacial valley left over by glaciers from
the last glaciation? The answer gathered by reading the literature is
rather simple, in fact too simple. The simplest model used is that
of a parabola, which is a curve with a U-shape, but this fact does
not mean that it is the correct profile. If x is a coordinate running
transversally to the valley and $y(x)$ is a function describing the transverse valley profile, the simple model is

$$y(x) = px^2 + qx,$$

where p and q are constants. Field work consists of taking measurements of glacial valley profiles and fitting the formula above
to the data, determining the numerical values of the parameters
p and q that give the best fit. This is really unsatisfactory for a
theoretically minded scientist. In fact, *any* profile which is smooth
and has a minimum is fitted by a parabola locally, that is, near its
minimum.[6] For this reason, this procedure conveys no information
when one wants to build models of how the glacier eroded the valley, to predict its final profile, and to compare the results with data.
Indeed, the use of parabolas to model glacial valley transverse profiles has been heavily criticized (*e.g.*, [Pattyn and Van Huele (1998)]),
but it is still thriving in geomorphology. A generalization consists of
using slightly more complicated functions to fit these profiles, such
as the simple power law

$$y(x) = ax^\alpha,$$

where a and α are constants and α now is not required to assume
the value 2, or to be an integer. The next generalization is using two
of these power law functions, one for each side of the valley,

$$y(x) = \begin{cases} ax^\alpha & \text{if } -x_1 \leq x < 0, \\ bx^\beta & \text{if } 0 \leq x < x_2, \end{cases}$$

---

[6]More precisely, for a smooth function $f(x)$ with a minimum at $x = 0$, the Taylor
expansion $f(x) = f(0) + \frac{f''(0)}{2}x^2 + \cdots$ always holds.

where a, b, α, and β are constants and the valley cross-section is spanned by the range $-x_1 \leq x \leq x_2$ of the transverse coordinate x. Again, this is mere data-fitting without physical or geological insight.

There are, however, detailed studies attempting to model the glacial erosion process and to follow it over tens of thousand of years. This process is complicated and its main features cannot be enclosed in one simple equation, belonging instead to the realm of numerical computation. Further, the equations describing the process contain several ingredients: one must assume a model of friction and boundary conditions describing the initial topography and the valley profile, and then model the stresses in the ice along the valley walls. The system of mathematical equations describing this process cannot be solved exactly by humans and computer calculations must be performed to evolve the valley model from a chosen initial state to its final configuration. This work has started recently [Seddik, Greve, and Sugiyama (2009); Yang and Shi (2015)]. The conclusions are that the erosion of the valley results from the interplay of two different effects due to the glacier, that is, widening and deepening of the valley [Seddik, Greve, and Sugiyama (2009)]. Different assumptions on the details of the model used to perform the calculations may emphasize one aspect over the other, which means that many detailed studies will have to be carried out before a "standard model" which is, relatively speaking, simple and reliable is finally formulated.

A clever alternative to this detailed work consists of skipping the modelling of the intermediate erosion process with all its details, and trying to determine the final glacial valley profile. In order to do that, one postulates that a certain physical quantity related to the system, or the process, is maximum or minimum in the final configuration. This procedure is common in mathematics and physics and is expressed by a *variational principle* [Goldstein (1980); Weber and Arfken (2004); Moiseiwitsch (1966)]. For example, suppose that you hang a piece of climbing rope from two points, unavoidably shaking it in the process. What goes on (some transient wave) between the initial state of the rope and its final configuration when it comes to rest is complicated. However, by assuming that the energy, which in this case, is just the gravitational potential energy of the rope, is minimum in the final configuration of equilibrium, we obtain a profile that corresponds to the actual one reached by the rope [Goldstein (1980)]. This curve is called a *catenary* and can be seen also in the shape of a wet, heavy spider web hanging from two points during your early morning start for a climb. The name comes from the Latin word *catena*, which means chain, since the curve is the shape of a long ideal chain suspended at two points. Any flexible wire, rope, or long chain will sag under its own weight and the shape of the curve that it describes in an (x, y) vertical plane is approximated by a catenary. The equation of a catenary is simply

$$y(x) = A \cosh x \equiv A \left(\frac{e^x + e^{-x}}{2} \right),$$

where A is a constant. A catenary with amplitude $A = 1$ is shown in Fig. 4.8. This curve is distinct from a parabola.

The mathematical process to obtain the catenary curve consists of writing down the potential energy of the rope, which depends on its shape, and of varying theoretically the shape of the curve among all the possible curves which connect the two endpoints, which are

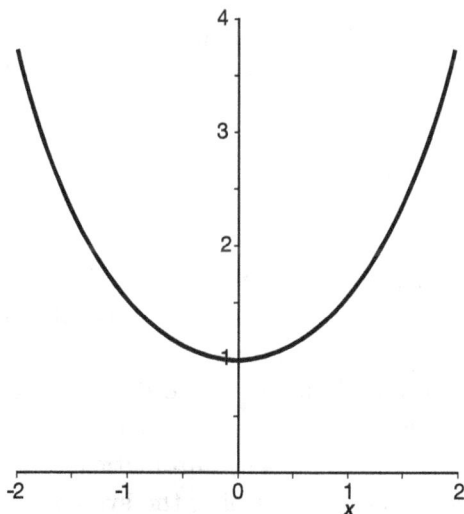

Fig. 4.8   A catenary curve.

kept fixed. The profile that minimizes the energy gives the correct catenary shape (see Appendix C for details). This procedure calls upon the *variational principle* and is used in many areas of mathematics and physics, where a certain quantity related to the system under examination is extremized (that is, minimized or maximized) to obtain the actual configuration of that system [Goldstein (1980); Weber and Arfken (2004); Moiseiwitsch (1966)]. Say that the quantity that one wants to extremize is given by an integral

$$J = \int_{x_1}^{x_2} dx\, L\left(y(x), \frac{dy}{dx}, x\right),$$

where $y(x)$ is an (unknown) regular function of the independent variable x and dy/dx is its derivative. The physics, or the geometry, of the problem is contained in the function L, called the *Lagrangian*, which is the kinetic energy T of the system minus its potential energy V:

$$L\left(y(x), \frac{dy}{dx}, x\right) = T - V.$$

It can be shown that extremizing the integral J, while keeping the endpoints x_1, x_2 fixed, that is, setting its variation $\delta J = 0$, always produces the equation [Goldstein (1980); Weber and Arfken (2004); Moiseiwitsch (1966)]

$$\frac{d}{dx}\left(\frac{\partial L}{\partial(dy/dx)}\right) - \frac{\partial L}{\partial y} = 0, \tag{4.6}$$

called *Euler–Lagrange equation*, which describes the motion, or the geometry, of the system. This procedure works like a well lubricated machine: if there is no dissipation (friction) in the system, then one can introduce a potential energy, write down kinetic and potential energy, the Lagrangian $L = T - V$ and the integral J, and then the Euler–Lagrange equation (4.6) automatically gives the equation of motion, or the equation describing the system [Goldstein (1980); Weber and Arfken (2004); Moiseiwitsch (1966)]. For example, for a particle with mass m and position $s(t)$ moving in one dimension under a conservative force $F = -dV/ds$, one can write down the potential energy and the Lagrangian and then the Euler–Lagrange equation (4.6) almost magically reproduces Newton's second law $m\, d^2s/dt^2 = F$.

The variational principle method can be applied to the problem of computing the transverse profile of glacial valleys [Hirano and Aniya (1988)]. In the case of glaciers, the quantity that is extremized is the friction force between the ice and the valley walls. For technical reasons [Morgan (2005)], one must also impose that the volume of ice going through the valley cross section remains constant [Morgan (2005)]. Early attempts to solve this problem obtained exactly a catenary curve as the cross-section profile, but this turned out to be incorrect. The best procedure thus far obtains a rather complicated differential equation describing the cross-section profile $y(x)$, but fortunately this equation has been solved exactly [Morgan (2005); Chen, Gibbons, and Yang (2015a)].

Let $y(x)$ be the thickness of the ice at transverse coordinate x and let us look for a smooth symmetric thickness profile $y(x)$ defined on the interval $-x_0 \leq x \leq x_0$ (with $x_0 > 0$), with $y'(0) = 0$ and $y(x) > 0$ for $|x| \leq x_0$. The variational principle for this problem takes the form [Morgan (2005)]

$$\delta J = \delta \int_{-x_0}^{+x_0} dx \left[y\sqrt{1 + y'^2} - \lambda y(x) \right] = 0,$$

where λ is a constant (*Lagrange multiplier*) and a prime denotes differentiation with respect to x. One can check that the Euler–Lagrange equation (4.6) then gives the differential equation expressing the extremization of friction subject to the constraint that the cross-sectional area $\int_{-x_0}^{+x_0} dx\, y(x)$ be constant:

$$\left(\frac{y'}{y} \right)^2 = \frac{1}{(\lambda y - C)^2} - \frac{1}{y^2}, \tag{4.7}$$

where C and λ are constants with $C > 0$ and $\lambda > 1$ required for the solution to make physical sense [Morgan (2005)]. The analytical solutions of this equation [Morgan (2005); Chen, Gibbons, and Yang (2015a)] do indeed represent U-shaped profiles. They are given by

$$\left(\lambda^2 - 1\right) |x| = C\lambda\sqrt{1 - w^2} + C\arccos w,$$

where

$$w = \left(\frac{\lambda^2 - 1}{C}\right) y - \lambda$$

and

$$-\frac{1}{\lambda} \leq w \leq 1,$$

$$\frac{C}{\lambda} \leq y \leq \frac{C}{\lambda - 1}.$$

In spite of the idea of the variational principle for this problem having been introduced in the 1980s, there is still work left to do in this area of research, as well as in the numerical modelling of the detailed erosion process forming the characteristic U-shaped glacial valleys. This thought should be of some consolation for those who are both mountaineers and scientists when their legs will no longer be able to carry them to the high peaks.

4.7   The universe in a glacial valley

Now comes a surprise: alert mathematicians [Chen, Gibbons, and Yang (2015a,b)] observed that the differential equation (4.7) ruling the ice thickness $y(x)$ that once occupied a U-shaped glacial valley (Sec. 4.6), is formally the same as a certain equation of cosmology (the Friedmann equation) which governs the evolution of the universe in the large under certain conditions [Liddle (2003)]. Any two points in our expanding universe have a relative distance d that increases with time. If d_0 is their initial distance at a certain time t_0, their distance at a later time $t > t_0$ will be $d(t) = d_0 a(t)$, where $a(t)$ (called *scale factor*) is a function of time which is universal, in the sense that it is the same for *all* pairs of points in three-dimensional space. Exchange x (the transverse coordinate in the glacial valley) with the cosmic time t and $y(x)$ with the scale factor $a(t)$, and the equation satisfied by these two functions is formally the same, provided that the universe is filled with some exotic form of matter. Now, the universe is really exotic, much more exotic than glaciers and mountains. We have known since the work of Edwin Hubble in the 1920s that the universe expands, but in 1998 the astonishing discovery was made that this expansion is accelerated and that this acceleration began only relatively recently in the cosmic history [Riess *et al.* (1998); Perlmutter *et al.* (1998); Riess *et al.* (1999); Perlmutter *et al.* (1999); Riess *et al.* (2001); Tonry *et al.* (2003); Knop *et al.* (2003); Riess *et al.* (2004); Barris *et al.* (2004)]. Before 1998 scientists believed that the universe originated in a Big Bang and, after this initial violent push, it expanded but was decelerated by its own gravity [Weinberg (1972); Misner, Thorne, and Wheeler (1973); Wald (1984)]. This would be analogous to throwing a rock vertically up in the air: the rock usually comes down on our heads. If we replace the rock with a rocket and we give it an initial velocity upwards of at least 11.2 km/s (*escape velocity*), it will escape the gravitational field of the Earth. But, once the rocket engines are turned off, whether it falls down, escapes, or barely makes it to infinity, it always decelerates until either it turns around, or all the way to infinity. The fact that the universe accelerates means instead that it is permeated by a very exotic form of energy (it has negative pressure) which is repulsive and dominates over the gravitational attraction of ordinary matter, which wants to decelerate its expansion. We do not know what this exotic form of energy, called *dark energy*, is. Various hypothetical forms of dark energy are being studied by theorists and probed by astronomers [Amendola and Tsujikawa (2010)]. Some of the dark energy models with the most exotic, even outrageous, properties are those for which the analogy with glacial valleys holds, that is, those for which the ruling differential equation (*Friedmann equation*) is analogous to the equation (4.7) for glacial valley transverse profiles derived with a variational principle. In physics and in engineering the advantage of an analogy in which two different physical systems satisfy the same formal equation, but the variables involved have very different physical meanings, is that one can learn about one system from knowledge of the other. Or perhaps the properties of a system that cannot be assembled and controlled in a lab, for example the universe itself, can be learned by studying a different, but analogous system which is smaller and more convenient to handle. In our case, it is very entertaining to draw a parallel between the glacial mechanism carving U-shaped valleys and the dynamics of a cosmos permeated by a form of very exotic dark energy [Faraoni and Cardini (2017)].

Let us look at this coincidence in some detail. The scale factor $a(t)$ of the universe must satisfy the *Friedmann equation*

$$H^2 \equiv \left(\frac{\dot{a}}{a}\right)^2 = \frac{8\pi G}{3}\rho - \frac{K}{a^2}, \tag{4.8}$$

where an overdot denotes differentiation with respect to the time t, $\rho(t)$ is the energy density of the cosmic fluid permeating the universe, and G is Newton's constant. The *curvature index K* is a constant normalized to 0 or ± 1 which discriminates between three possibilities: open universe ($K = -1$, with infinite volume, expanding forever), closed universe ($K = +1$, with finite volume, which expands to a maximum size and then contracts and collapses), or critically open universe ($K = 0$, the boundary between the two previous situations — this is still an open infinite universe which expands forever). The expansion rate $H \equiv \dot{a}/a$ is called *Hubble parameter*. The scale factor must satisfy also the acceleration equation

$$\frac{\ddot{a}}{a} = -\frac{4\pi G}{3}(\rho + 3P), \tag{4.9}$$

where $P(t)$ is the pressure of the cosmic fluid permeating the universe. Another equation can be derived from the Friedmann and the acceleration equations,

$$\dot{\rho} + 3H(P + \rho) = 0. \tag{4.10}$$

This equation expresses the conservation of energy for the cosmic fluid.

The equation describing the transverse profile of a glacial valley is Eq. (4.7) of the previous section,

$$\left(\frac{y'}{y}\right)^2 = \frac{1}{(\lambda y - C)^2} - \frac{1}{y^2},$$

where $y(x)$ describes the ice thickness as a function of the transverse coordinate x and λ and C are constants with $\lambda > 1$ and $C > 0$. A quick inspection shows that this equation is analogous to the Friedmann equation (4.8) of cosmology with $K = +1$ describing a closed universe, if we do the exchange

$$t \longleftrightarrow x,$$

$$a(t) \longleftrightarrow y(x),$$

$$\dot{a} \equiv \frac{da}{dt} \longleftrightarrow y' \equiv \frac{dy}{dx}.$$

The cosmological analogue of the glacier equation (4.7) is

$$\frac{\dot{a}^2}{a^2} = \frac{1}{(\lambda a - C)^2} - \frac{1}{a^2},$$

which can be rewritten in a form more familiar to cosmologists as

$$\frac{\dot{a}^2}{a^2} = \frac{8\pi G}{3} \frac{\rho_0}{(a - a_0)^2} - \frac{1}{a^2},$$

where

$$\rho_0 = \frac{3}{8\pi G \lambda^2} > 0, \qquad a_0 = \frac{C}{\lambda} > 0.$$

Then the energy density in the cosmic analog of the glacier valley must have the dependence on the scale factor

$$\rho(a) = \frac{\rho_0}{(a - a_0)^2} > 0.$$

For the analogy to be valid, we must impose one of the remaining two cosmological equations, then the third one will automatically be satisfied because it is not independent of the other two. Let us impose Eq. (4.10). This condition gives immediately the pressure of the cosmic fluid in the cosmic analog of the glacial valley

$$P(a) = \frac{2\rho_0 a}{3 (a - a_0)^3} - \frac{\rho_0}{(a - a_0)^2},$$

or

$$P(\rho) = \frac{-\rho}{3} \pm \frac{2a_0}{3\sqrt{\rho_0}} \rho^{3/2},$$

where the upper sign applies if $a > a_0$ and the lower one if $a < a_0$. The relation between P (in units of ρ_0) and ρ (also in units of ρ_0) is plotted in Fig. 4.9.

What kind of cosmic fluid satisfies this relation between P and ρ (called *equation of state* and characteristic of the type of fluid)? None that is known on Earth, but it reproduces some speculative form of dark energy that has been contemplated by theoretical cosmologists [Barrow (2004); Nojiri and Odintsov (2004); Štefančić (2005); Nojiri *et al.* (2005); Capozziello *et al.* (2006); Ananda and Bruni (2006a,b);

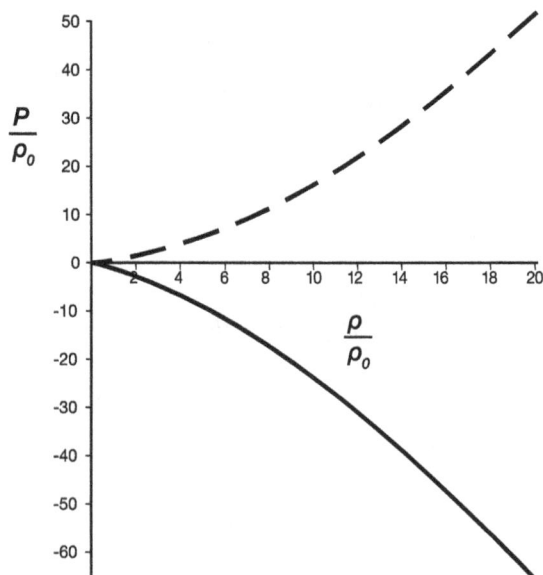

Fig. 4.9 The pressure as a function of the energy density ρ for negative sign (solid curve) and for positive sign (dashed curve) in Eq. (4.7), for $a_0 = 1$.

Silva e Costa (2009); Frampton *et al.* (2011); Borowiec *et al.* (2012); Chen, Gibbons, and Yang (2015a,b); Bouhmadi-Lopez *et al.* (2015); Borowiec *et al.* (2016)]. A universe with $a(t) > a_0$ at all times constitutes a meaningful analogue of a transverse glacial valley profile. In cosmology, this fluid would be very exotic because of a peculiarity: one would have a spacetime singularity, where the curvature of spacetime becomes infinite, while the density ρ and the pressure P of the cosmic fluid remain well behaved. Further, this singularity would occur at a finite time in the future evolution of the universe, not at an infinite time, as for a universe expanding forever. The other situation which is mathematically possible, in which $a < a_0$, is not a meaningful analogue of a glacial valley [Faraoni and Cardini (2017)].

Here the cosmic scale factor $a(t)$ corresponds to the ice thickness $y(x)$, which is maximum at $x = 0$ (corresponding to $t = 0$) and minimum at the valley boundaries $x = \pm x_0$ (corresponding to $t = t_0$). Of course, the analogy is purely formal, but mathematical analogies are quite powerful and should not be overlooked. For example, one can describe a pendulum or any linear mechanical oscillator with a simple linear electrical circuit and *vice-versa* because these two very different physical systems, one mechanical and the other electrical, obey the same mathematical equation [Goldstein (1980)]. Indeed, before computers were available, powerful, and cheap, there were people specializing in building electric circuit analogues of "real" physical systems that would be impossible or prohibitive to build in the lab. For example, one cannot build a universe in a lab to study its behaviour, but one can try to build an analogue circuit obeying the same equations. Computer simulations and the enormous increase in computing power since the 1960s have made these electrical analogues obsolete since one can now solve numerically the mathematical equations relatively easily and one can often run computer simulations to study the behaviour of a physical system. It is nevertheless interesting that glacial valleys are natural analogues of hypothetical (maybe even real?) universes filled with extremely exotic dark energy fluids.

Chapter 5

Heat, cold, and air

5.1 Introduction

Heat, cold, and temperature changes are fundamental aspects of life, but especially of outdoors life in the harsh mountain environment. On average, the surface of the Earth receives a flux density of radiation (called *solar constant*) $S = 1.370 \cdot 10^3$ W/m^2. Differential heating by solar radiation, temperature gradients, and the subsequent heat exchanges are the fundamental causes of many natural phenomena upon which the success of a climb, or even the life of the mountaineer depend crucially. Think of air movement and of winds created by temperature gradients, of the transport of snow by these winds which creates snow pillows on lee slopes, treacherous cornices along ridges, and flimsy snow bridges over crevasses; of the summer thunderstorms which we fear when we are on a rock wall or on a summit or a long ridge from which we can not escape quickly; of sudden snowstorms changing an easy scramble into a little epic (Fig. 5.1); or of the challenge of keeping warm at high elevation, of the radiative cooling on a clear, cold night, and of the need to choose campsites wisely.

Changes in insolation and in the amount of radiation energy received from the sun mark the day and night cycle and the alternation of the seasons, which are more pronounced in the mountain environment than at lower elevations. Heat transfer concerns us directly

Fig. 5.1   Less than perfect summer weather in the Karwendel, Austria.

as mountain climbers because of the need to regulate the temperature of the human body and to maintain it within a narrow range, with a penalty of frostbite and potential loss of limbs for failing to do so. We are concerned about atmospheric pressure and especially about its changes signaling that the weather is improving or deteriorating, and about how fast these changes happen, which is usually related to the magnitude of these changes. Reading signs in the sky, such as sundogs and halos around the moon, and weather forecasting are very useful skills when we are on a long trip in the mountains. I find myself using often the barometer in my watch which, incidentally, is based on a semiconductor device that changes its conductivity as the atmospheric pressure changes.[1]

Phase changes are particularly evident and important in alpine regions, causing freezing and the stabilization of a snowpack, melting

---

[1]The design of these solid state devices is entirely based on quantum mechanics.

Fig. 5.2 The snowline separating ablation and accumulation zones is clearly visible (Sesvenna Glacier, Switzerland).

and the swelling of small streams crossed easily in the morning which change to raging torrents at the end of the day, the snowline on a glacier receding as the season progresses (Fig. 5.2), evaporation, condensation, and sublimation. Related effects include snowfall, rainfall, clouds rising, fog, whiteouts and getting lost, frost and cold fingers, and thunderstorms and the very concrete risk of being struck by lightning up high. Ablation from solar radiation creates suncups in snowfields and penitentes on high dry glaciers. The variety of alpine phenomena related to thermal physics is enormous.

5.2 Why is it cold up there?

It is common knowledge that it is generally colder in the mountains than at lower elevations. Traditionally, during hot summer days, people who could would search relief in the hills from the heat of the plains. The temperature decreases about 6 degrees Celsius every 1000 meters of elevation gained and the humidity in the air condenses, forming clouds (the exception is when there is thermal inversion, which occurs especially in winter). Why is it cold up there?

Qualitatively, if air was an ideal gas, and let's assume it is for simplicity, and if a parcel of air ascends without exchanging heat with its surroundings (in slang, it expands *adiabatically* as it goes up), its temperature increases when the pressure increases and it decreases if the pressure decreases (details later). Since, going up higher, the air density and pressure decrease, also the temperature does.

A simple model of this process is a classic exercise from university textbooks [Carter (2001); Fermi (1956); Sears and Salinger (1975); Zemansky and Dittman (1997)], which is proposed in the following. Assume that air is an ideal gas, that is, its molecules are far apart, they only feel each other when they collide, and then bounce off each other perfectly elastically. Consider a parcel of air with the shape of a horizontal parallelepiped (a box), in equilibrium, as in Fig. 5.3. Equilibrium is achieved because of the balance between its weight and the pressure gradient between the top, where the pressure pushes

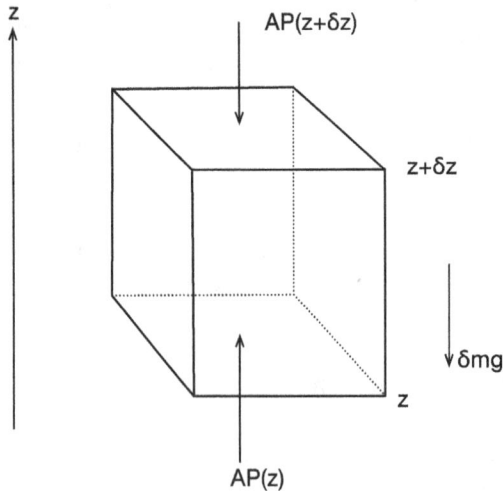

Fig. 5.3 A fluid parcel of infinitesimal volume $\delta V = A\delta z$ in hydrostatic equilibrium.

the box down, and the bottom where a larger pressure pushes it up. At equilibrium there is no net force on the air parcel, which does not accelerate because force = mass times acceleration according to Newton, and here the force is zero. We say that the parcel is in hydrostatic equilibrium. This hydrostatic equilibrium is described by a differential equation which involves the (vertical) pressure gradient, which is derived in Appendix B. If z measures the elevation along a vertical axis starting at sea level and pointing up, the equation of hydrostatic equilibrium is

$$\frac{dP}{dz} = -\rho g, \tag{5.1}$$

where ρ is the mass density of air, P is the pressure, and g is the (constant) acceleration of gravity.

At this point, the ingredients to use are two laws that express the physics of an ideal gas. The first ingredient is the ideal gas law, or ideal gas equation of state

$$PV = \frac{m}{M} RT, \tag{5.2}$$

where m is the mass of the gas, M is its atomic mass, and R is the universal gas constant. The second ingredient is the law for adiabatic transformations of an ideal gas

$$TP^{\frac{1-\gamma}{\gamma}} = \text{constant}, \tag{5.3}$$

where the *adiabatic index* $\gamma > 1$ is a constant characteristic of the particular gas considered[2] [Carter (2001); Fermi (1956); Sears and Salinger (1975); Zemansky and Dittman (1997)]. Combining Eqs. (5.2) and (5.1), one obtains

$$\frac{T}{P} \frac{dP}{dz} = -\frac{gM}{R}. \tag{5.4}$$

---

[2]The adiabatic index is the ratio of the specific heats of the gas at constant volume and at constant pressure, $\gamma = c_P/c_V$. It is $\gamma > 1$ because it is always $c_P > c_V$.

Now one uses the law for adiabatic transformations of an ideal gas (5.3) to eliminate the pressure P from the last equation, obtaining

$$\frac{dT}{dz} = -\frac{(\gamma - 1)}{\gamma} \frac{gM}{R}.$$

Since $\gamma > 1$, this temperature gradient is negative, meaning that the temperature decreases with elevation. Let us insert the known numerical values for air, modelled as a gas with an average molecule: $\gamma = 7/5$ for an ideal gas composed of diatomic molecules [Carter (2001); Fermi (1956); Sears and Salinger (1975); Zemansky and Dittman (1997)], $g = 9.81\,\text{m/s}^2$, $M = 28.88$ atomic mass units, $R = 8.214\,\frac{\text{J}}{\text{mol·K}}$, then one obtains the temperature gradient

$$\frac{dT}{dz} = -9.8\,\frac{\text{K}}{\text{km}} \tag{5.5}$$

which is, in absolute value, quite a bit larger than the observed $6\,\text{K/km}$ (keep in mind that *intervals* of Kelvin temperature are the same as intervals of Celsius temperature, although Celsius and Kelvin temperatures differ). This discrepancy is due to the fact that atmospheric air contains water vapour, which has thermal inertia and mitigates the temperature gradient. Assuming constant air composition, Eq. (5.5) is integrated vertically, giving a linear decrease of temperature with elevation,

$$T(z) = T_0 - \alpha z,$$

with $\alpha = \left(\frac{\gamma-1}{\gamma}\right)\frac{gM}{R}$ a constant.

We can consider a region of size $z \ll \alpha^{-1}$ in which the temperature is approximately constant and use again Eq. (5.4), which can be re-written as

$$\frac{d}{dz}\left[\ln\left(\frac{P}{P_0}\right)\right] = -\frac{gM}{RT},$$

where P_0 is a constant. This equation is easily integrated when the right hand side is constant, producing

$$P = P_0\,e^{-z/H}, \quad H = \frac{RT}{gM}, \tag{5.6}$$

known as the *barometric formula*. Since a given mass m of gas at temperature T, pressure P, and volume V has density

$$\rho = \frac{m}{V} = P\frac{M}{RT}$$

(proportional to the pressure), as follows from the ideal gas law (5.2), the air density is obtained as

$$\rho = \rho_0\, e^{-z/H}. \tag{5.7}$$

The constants $P_0 = P(0)$ and $\rho_0 = \rho(0) = \frac{P_0 M}{RT}$ have the meaning of pressure and density at the level $z = 0$, respectively. Therefore, over a region in which the temperature is approximately constant, the air density and pressure decrease exponentially with elevation. No wonder that there is little air to breath or to fly rescue helicopters at 8000 m of elevation. The constant H has the meaning of vertical length scale over which the pressure and density vary. If the elevation increases by H, the air density and pressure decrease by a factor $e \simeq 2.7$. Using an average temperature $T = 250$ K $= -23°$C, one obtains $H = 7.4$ km for the atmosphere, which means that in this model the pressure and density of air are reduced to half at an elevation of approximately 5000 m above sea level.

In actual fact, the temperature is not constant and the exponential law is not strictly valid. By using $T(z) = T_0 - \alpha z$, one finds instead

$$P(z) = P_*\, (T_0 - \alpha z)^{\frac{\gamma}{\gamma-1}}$$

with P_* an integration constant (not the pressure at sea level), while

$$\rho(z) = \rho_*\, (T_0 - \alpha z)^{\frac{1}{\gamma-1}}$$

and $\rho_* = P_* M/R$. The qualitative behaviour of $P(z)$ and $\rho(z)$ in this case is plotted in Fig. 5.4 for $\gamma = 7/5$.

5.3   The dress code

The atmosphere is less dense in the mountains and, in good weather conditions, it contains less water vapour than in the valleys or at

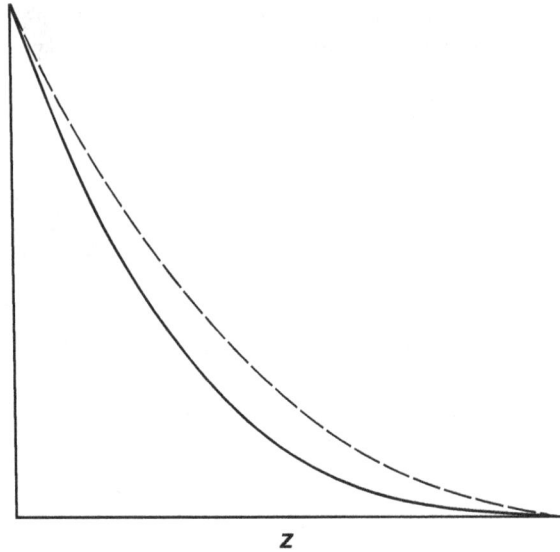

Fig. 5.4   The qualitative behaviour of the pressure $P(z)$ (solid curve) and density $\rho(z)$ (dashed curve) when $T(z)$ decreases linearly.

sea level. Water vapour contributes greatly to thermal inertia and mitigates temperature variations. In dry conditions, the air absorbs and scatters less of the solar radiation crossing the atmosphere than in humid conditions, and temperature variations in the mountains are much more pronounced than at lower elevations. A similar phenomenon happens in a desert: there is scarce water vapour in the air, which would cause thermal inertia and smooth out the variations of temperature. In a desert the temperature rises suddenly when the sun comes up and drops abruptly when the sun goes down. In the mountains, in addition to scarce water vapour, the air is also thinner and one may be freezing overnight while waiting to start a climb and be very hot, thirsty, and dried out the next morning while high up on a rock wall hit by direct sunlight and by the reflection from the snow and ice below. Because of the low density of air and of its low water vapour content, there are also strong temperature differences between sunny and shady areas. If we plan to spend the day on a rock route at high elevation, it is better to know if it stays in the shade or not.

These sometimes extreme temperature variations should be kept in mind when planning the way we dress for the mountains, or when we buy a sleeping bag or other gear. Another rule of thumb to apply when choosing clothing items is that the temperature decreases approximately 6 degrees Celsius every 1000 meters of elevation (Sec. 5.2). In general, it is convenient to adopt the multi-layer scheme: instead of a single heavy layer, one uses multiple lighter layers which can be stripped or added as needed. The thin layers of air trapped between layers of clothing add to the insulation. In addition to flexibility, this scheme has the advantage of not relying on a single heavy item of clothing which can become wet or get lost.

5.4 Conserving heat

On a cold mountain, conserving heat and preventing hypothermia is absolutely necessary and requires some skill. According to the physicist, heat is transferred by three mechanisms: *conduction, convection,* and *radiation.*

Conduction requires direct contact between the body which loses heat and the one which gains it. For example, we lose heat by conduction when we sit on the ground without insulation, or when we hold a metal ice axe without gloves, or even with gloves. At low temperature, moisture in the skin can freeze on a metal surface. Leading an ice climb, my friend John took an ice screw in his mouth for a moment in order not to drop it while he was adjusting some gear on his harness. A piece of skin lost from his lip when he took the screw back in his gloved hand alerted him about the sudden phase change of the moisture in his lip.

Convection is heat transfer due to the motion of a fluid which removes heat from a warm body and transfers it to its surroundings. For example, moving air cools by convection. Still, confined air is instead a good thermal insulator. In fact, birds fluff up their feathers to keep warm, and we use down jackets which trap air in between the feathers but wind, drafts, and currents remove heat quickly by convection. This phenomenon is particularly evident at low temperatures, hence the introduction of the windchill factor estimating an effective temperature in the presence of wind. The solution to this convective cooling is, of course, to get out of the wind and to wear a windproof outer layer.

Radiation transfers heat without contact with solid or fluid. Every object emits electromagnetic radiation, the more so when its temperature is higher. Even without touching cold objects and in still air, we lose heat by emitting infrared radiation into cooler surroundings. The solution is, of course, to cover up in order to limit the heat loss. To prevent heat loss by radiation, wearing headgear is particularly important because relatively large amounts of heat are lost through the head and neck.

In practice, one of the three mechanisms of heat transfer may be dominant and become by far the most important, or two or three mechanisms may operate at the same time with comparable efficiencies. Once I climbed the Eisenhower Tower in the Canadian Rockies, by all means not a difficult route, but because of a combination of getting off route and backtracking halfway up, being a little too slow on the climb, and a small injury to my climbing partner from a falling rock on a rappel, we were benighted. We were on a huge ledge from which only 150 meters of easy scrambling separated us from the trail, but one has to find the exact spot to descend and then one must stay on route or else end up on a steep wall with plenty of rotten rock. Our headlights were not sufficiently powerful to find the descent route and, after a couple of attempts ending on said rotten rock, we set out to spend the night on the big ledge. It wasn't cold by any mountaineering standards but it was still an unplanned bivouac and heat conservation was a must. There was a miniature pine tree on the ledge and we laid down by it to cut the breeze (convection). Since we had climbed with two half-ropes, we had a rope each and we laid it down on the mossiest spot we could find, together with our rock shoes, slings, and webbings from quickdraws, stripped of the metal carabiners of course, to get some insulation from the ground (conduction). A large garbage bag[3] that each of us always carries

---

[3]The garbage bag, this amazing multi-use piece of gear is light, cheap, and does not need a specialized store to acquire. Better if it is orange instead of black, so

on a climb (unless we carry a proper bivy sac, that is) kept out the abundant dew of the night and kept some warm air around our bodies (convection). A warm hat and gloves, which are light and fit even in the small backpack carried on a technical rock route completed the job (radiation). The next morning we scrambled down to the trail in daylight, hiked back to our tent, and drove somewhere for breakfast and lots of warm strong coffee. I cannot say that I slept a lot that night but it wasn't too bad, and it would have been a lot worse had we not found ways to minimize the heat transfer from our bodies to the surrounding environment.

In addition to the physics textbook mechanisms of heat transfer, we should mention *evapotranspiration*. By now we know that turning liquid water into vapour requires a relatively large amount of heat. A wet body from which water evaporates provides this heat, which is lost, and the body cools. In low temperatures, immersion in water can be critical, but also wearing wet clothing can range from uncomfortable to dangerous. Of course, the solution is to wear dry clothes instead of wet ones, which has led to the introduction of polar fleece and synthetic fiber clothing which removes moisture to the outside and minimizes dampness. Cotton clothing is practically banned from high or cold mountains since saturated cotton has a thermal conductivity near that of water and heat loss by evaporation becomes a big problem — we wear fleece or wool instead. Wool retains warmth even when wet, but synthetic fabrics are lighter and preferable in practice. Waterproof, breathable clothing is the choice but, if it is windy, it must be supplemented by a vapour barrier, a windproof layer. In fact, polar fleece is almost too efficient, in the sense that in cold weather moisture is removed efficiently but evaporation on the outside layer may lower the temperature too much and it is preferable to surround it with a vapour barrier confining the warm air near the body [Twight (1999)]. Needless to say, also respiration is a cause of heat loss since we inhale cold air and exhale warm air. In extreme conditions a handkerchief or face mask can keep warmer

---

it can be spotted easily from a helicopter in the unfortunate circumstance that rescue is needed. Never leave it behind!

air near one's face, but then the water vapour contained in warmer air from the body freezes. Beards full of icicles adorn high altitude mountaineers and polar explorers and make for telling pictures.

Instinctive tricks to conserve heat include making oneself smaller, kneeling down, or two or more people huddling together. The physical explanation of this behaviour is simple. Model a body as a sphere of radius R: heat is lost through the surface in contact with the cold environment (conduction) or exposed to convection. The surface area of the body-sphere is $A = 4\pi R^2$. Heat is generated in the bulk of the body, so the amount of heat generated is proportional to its volume, which is $V = 4\pi R^3/3$ for this body-sphere. In order to conserve heat, one wants to minimize the surface-to-volume ratio, that is, decrease the surface area A, or increase the volume V, or both. This ratio,

$$\frac{A}{V} = \frac{4\pi R^2}{4\pi R^3/3} = \frac{3}{R},$$

is inversely proportional to the radius R. A large R (large body) means a small ratio A/V and smaller amounts of heat lost, while a small R means large $1/R$ and larger amounts of heat lost — a larger body conserves heat better than a smaller one. Small children ("spheres" with small R) cool and become hypothermic faster than a big adult, and this is something to keep in mind when taking children outdoors in cold weather. A big guy will have some advantage over a skinny short one as an arctic explorer.

In practice, we are not spheres and the spherical body above is just a toy model, but the principle learned still applies. To conserve heat we reduce the surface which loses heat by covering it with insulating clothes and by minimizing the skin surface exposed, limiting it to the face and even covering it up with a face mask. We seek shelter to stay out of the wind and to limit radiation losses. Wilderness survival courses and manuals are full of ingenious ways to make a shelter: in an emergency in the alpine it usually boils down to digging a snow cave or a trench in the snow, or using a crevasse as a shelter. If lucky, when caught out below treeline, the bottom of a conifer tree with the lower branches embedded in the snow can provide a shelter with minimal work. On high windy alpine terrain with no precipitation and no snow, like Mt. Aconcagua and many peaks of the Andes, one must resort to building stone walls to shelter from the wind as best as possible. These stone shelters are often built also by shepherds and climbers at lower elevations in the Alps and the Pyrenees.

In extreme cold people huddle together: bringing more bodies together reduces the surface area while increasing the total volume. The parts of body surfaces which are in contact with each other are now in the bulk of the bigger "body". In the spherical toy model above, this trick is equivalent to merging two or more spheres of smaller radii into one of larger radius. The ratio A/V scales as $1/R$ and the total amount of heat lost versus heat generated (per unit time) decreases. Many mountaineers and lost hikers have survived cold nights huddling together and some climbers have been found frozen to death on a summit, at the end of a challenging technical route, in that position meant to minimize heat loss during a sudden storm. And, of course, it is not just a human reaction: bears and other animals who spend the winter hibernating in a den naturally arrange their bodies to be in contact with each other. Japanese macaques are also known to huddle together in cold weather to conserve heat. It is an instinctive reaction and the simple model of a sphere catches it. The *Bergmann rule* of biology [Bergmann (1847)] states that warm-blooded animals reach bigger size in colder environments and is explained by the same thermal physics and the scaling[4] $A/V \simeq 3/R$.

In the case of hyperthermia, one has the opposite problem and wants to increase the heat loss from the body instead. Hyperthermia symptoms include excessive sweating, loss of body fluids and, in extreme situations, flushed, hot skin and heat stroke. Of course, one should avoid heavy exercise in hot environments, but this is not always possible in the outdoors. Cooling through immersion is

---

[4]Then why do elephants live in the plains of Africa instead of on top of Mt. Kilimanjaro, where it is cold? The answer may lie in the fact that their enormous ears dissipate heat very efficiently to compensate for the size of these animals.

desirable since convection by the cooler fluid removes heat, but also sponging, spraying, and the resulting evaporation achieve some effect. A humid environment prevents in large measure cooling by sweating and evaporation. Since warm air accomodates much more water vapour than cold air, hot climates feel "sticky" and uncomfortable.

5.5   Föhn, chinook, and bicycle pumps

What do föhn and chinook winds have in common with bicycle pumps? The föhn is a warm dry wind well known in Switzerland, Austria, and Central Europe as a "snow eater" because it can raise the temperature by as much as 30 degrees Celsius in a few hours and it causes the rapid decrease or even disappearance of the snowpack due to both melting and sublimation. It will certainly change the conditions on our alpine climb! Similar winds are the chinook in the Canadian Rockies and the bergwind in South Africa.

How does such a warm wind originate? The key to understanding its origin is the behaviour of gases which are compressed or expanded quickly. When we compress a gas so quickly that it does not have the time to exchange heat efficiently with its surroundings, it warms up. Think of air in a bicycle pump: when we pump quickly and repeatedly, the pump warms up. A process involving a gas which does not exchange heat with its surroundings is called *adiabatic*. It could be that the gas is in a thermally insulated container, or it could be that the gas is in thermal contact with its surroundings but it is compressed so fast that it has no time to exchange heat (adiabatic compression). Similarly, a gas that expands very quickly undergoes an adiabatic expansion. In our case, the gas is air: think, for simplicity, of a horizontal wind, carrying moisture, which finds a mountain slope in its path: the air is forced up the slope quickly (*orographic lifting*) and undergoes an adiabatic expansion, during which it cools to the point that the moisture condenses into clouds. On top of a mountain ridge, it is common to see clouds which clearly originate on one side of the ridge, while the other side is cloud-free. Other times, cool ascending air encounters humid air on the ridge which comes from the opposite side, and which condenses on the spot due to sudden cooling. Or, humid air which ascends and cools adiabatically condenses and causes rain, while the opposite side of the mountain ridge remains dry.

In the reverse situation, cool dry air flows down the slopes of a mountain range. The adiabatic compression heats this dry air, causing a föhn or a chinook. The temperature rise can be surprising and it can cause extensive snow melting. Because the air is so dry, it also causes intense sublimation and the air flow removes moisture near the surface of the snow, which facilitates sublimation. It is the nearly adiabatic compression moving down the slope that heats up the air. Just like in a bicycle pump.

5.6 Valley breeze and mountain breeze

We have already seen that the specific heat of water (the heat energy necessary to raise the temperature of 1 kg of water by $1°$ C) is unusually large in comparison with most other common substances. As a consequence, water has a much larger thermal inertia than other substances. Think of a green mountain valley with lots of water, coming from the peaks above, and full of woods, grass, and vegetation, in contrast with the bare rock walls of the mountains surrounding them. Because of its water content, this green valley's thermal inertia is much larger than that of the surrounding rock walls. This means that, early in the day, the rocks will warm up quickly while the valley stays cool for much longer. Warm air on the peaks expands and rises, due to Archimedes' principle, leaving its place to cooler air from the valley: we have a warm valley breeze flowing upslope in the morning (Fig. 5.5).

Vice-versa, during the night the rocks have cooled off quickly because of their small specific heat, while the valley retains some of the heat of the day because of its water content. The warm air in the valley is less dense than the air from the passes and peaks and it rises, being replaced by cooler air which flows down the slopes: we have a mountain breeze at night (Fig. 5.6).

The same qualitative explanation applies to sea breezes and land breezes in coastal areas. A disruption of the regular daily cycle of

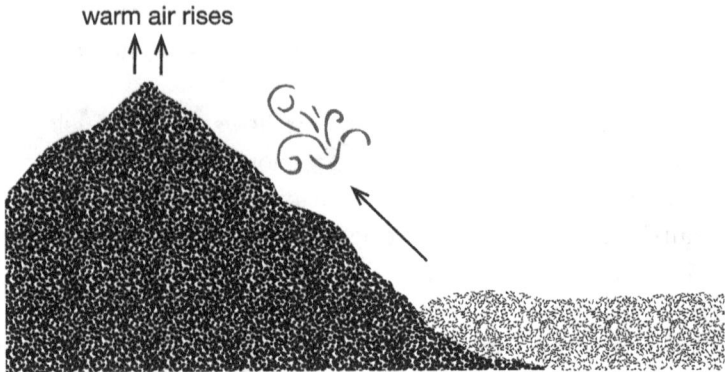

Fig. 5.5   Valley breeze in the morning.

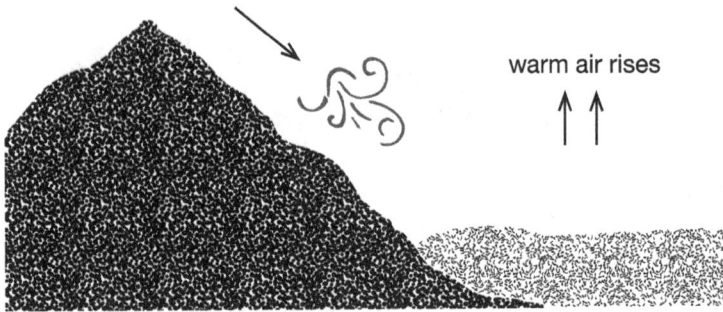

Fig. 5.6   Mountain breeze at night.

valley and mountain breezes may signal that the weather is changing and air masses are moving in.

The particular electric features of the H_2O molecule make water an unusual substance and that morning breeze was ultimately a consequence of microscopic molecular physics.

5.7   Clear nights are cold nights

If we bivouac or camp out on a mountain, or if we take off at one in the morning to climb a high peak, we are reminded that clear starry nights are beautiful but cold. Why is that?

An object warmer than its surroundings loses heat, which is transferred from the object to the surroundings. As we have already seen, there are three mechanisms of heat transfer: conduction, convection, and radiation, to which we add evapotranspiration if the "object" is a plant or an animal. Conduction occurs by direct contact when the atoms or molecules of the material transfer energy to atoms or molecules of another body, for example, we lose heat through conduction by touching the metal of an ice axe, or through boots that are not good, so the feet lose heat that is conducted away by metal crampons or snow. Convection occurs when masses of fluids carry warm air away from around the body, for example when wind sweeps away the warm air close to our bodies. Radiation is the emission of electromagnetic waves through air from the warm object to the colder surroundings. It happens even through a vacuum. Any object emits electromagnetic radiation with a characteristic spectrum which is peaked at a certain wavelength. A red hot piece of metal emits a spectrum peaked in the visible band while, at the temperatures characteristic of the human body, radiation is emitted in the infrared band and is invisible to the human eye. Some species of snakes, though, can sense the infrared. Also the Earth emits infrared radiation into outer space. If present, a cloud cover acts as a blanket which traps the infrared radiation and doesn't allow it to escape into space. Some infrared radiation emitted from the ground is then scattered back and does not escape freely to outer space. The effect is similar to that of a greenhouse, in which the transparent glass cover lets in short wavelength visible radiation during the day but is opaque to long wavelength infrared radiation, which is re-emitted by the plants inside the greenhouse. The principle is the same as in the greenhouse effect discovered by mountaineering pioneer John Tyndall [Tyndall (1872)] and causing global warming. Greenhouse gases in the atmosphere trap infrared radiation, increasing the average temperature of the planet. On a more local scale, a cloudless night on a glacier climb is going to be colder than one when a cloud cover is present (wind only makes it worse). Clear nights are cold nights.

Now the technical part: a body at absolute (or Kelvin) temperature T emitting electromagnetic radiation over the entire spectrum of frequencies ν emits more in certain frequency bands and less in others. A standard model describing how the energy of this radiation is distributed over all these frequencies is the *blackbody*. A blackbody can be constructed by removing a brick from the wall of a furnace so that a narrow pencil of radiation from the interior escapes without perturbing the equilibrium of the radiation remaining inside, and it can be analyzed. In practice, a blackbody is used to model almost any object emitting thermal radiation, although its temperature is not that of a furnace. The energy per unit volume and per unit frequency $u(\nu, T) = dE/dV\,d\nu$ at a certain frequency ν inside a blackbody at (absolute) temperature T is given by the renowned *Planck law*

$$u(\nu, T) = \frac{8\pi h\nu^3}{c^3}\,\frac{1}{e^{\frac{h\nu}{KT}} - 1}, \tag{5.8}$$

where h and K are known as the Planck constant and the Boltzmann constant, respectively, while c is the speed of light *in vacuo*. This law, discovered by the German physicist Max Planck in 1900, is famous because it marked the beginning of quantum mechanics [Griffiths (2005); Gasiorowicz (2003); Schiff (1968); Messiah (1961)]. At the end of the 1800s, calculations based on classical physics, made to model the well-known blackbody experimental curves were producing complete nonsense. Planck succeeded in calculating the correct law (5.8) only by postulating that the energy inside the blackbody cavity is emitted and absorbed in discrete packets by the atoms in the walls of the cavity. The energy of a single packet (or *quantum*) of radiation is $h\nu$, proportional to its frequency ν through the Planck constant h [Griffiths (2005); Gasiorowicz (2003); Schiff (1968); Messiah (1961)]. Note that some microscopic model of matter is necessary because the radiation inside the cavity is constantly absorbed and re-emitted by atoms in the walls of the cavity, and the electromagnetic radiation inside this cavity forms a *thermal bath* in a state of equilibrium.

The temperature T enters as a parameter in the function $u(\nu, T)$ in Eq. (5.8). The blackbody curves of the spectral energy density $u(\nu, T)$ for different temperatures form a one-parameter family with T as the parameter and do not intersect each other for $\nu > 0$.

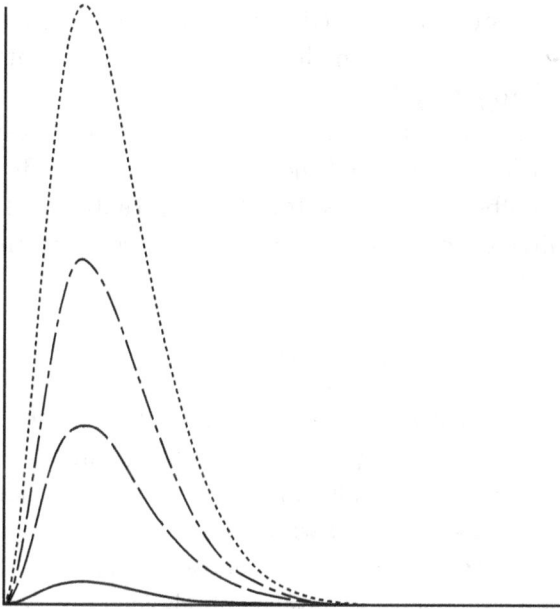

Fig. 5.7 Blackbody curves for temperatures T (solid), $2T$ (dashed), $2.5T$ (dash-dotted), and $3T$ (dotted).

These curves are plotted in Fig. 5.7 for temperatures $T, 2T, 2.5T$, and $3T$. For a given temperature, the total (*i.e.*, integrated over all frequencies) energy density in the blackbody cavity is the area under the curve.

Planck's discovery and the work of others which came soon thereafter, including Einstein, changed microscopic physics forever and made possible, for example, to build the semiconductor chips present in mountaineers' watches, cellphones, altimeters, and GPS units. All this goes hand in hand with the explanation of why clear nights are cold nights. Thank you, Professor Planck.

5.8 Archimedes in the tent

Most substances contract when they cool, and air is no exception. The density ρ of air is its mass m divided by its volume V,

$$\rho = \frac{m}{V}.$$

Therefore, when air contracts (that is, when the same mass m of air goes from a volume V to a smaller volume V'), it density *increases*, or goes from ρ to $\rho' > \rho$ because the denominator V decreases while the numerator m stays the same. Already in the second century B.C., Archimedes of Syracuse stated the famous principle of fluid mechanics saying that a body immersed totally or partially in a fluid receives an upward-directed push with force intensity equal to the weight of the volume of fluid displaced,

$$F_{\text{up}} = \rho_{\text{fluid}} V_{\text{displaced}} g.$$

Archimedes' principle applies also to a fluid immersed in another fluid, like hot water in cold water, or warm air in cold air. A consequence is that denser fluids sink, while warmer fluids rise. This principle applies also to gases, which is why a hot air balloon lifts.

Back to mountaineering: when choosing a site for our tent up on a mountain, we may not have much choice. But other times in alpine terrain we have meadows, little valleys, bumps, mini-ridges, bowls, and some choice between these options. And, when camping on snow, we do not need to worry about rocks when looking for a flat spot. Then it is useful to keep in mind that cold air is heavier than warm air and it sits in bowls, and you will probably feel the cold more at night when you are not moving. It is true that a bowl may provide shelter from winds — plus, a tent flapping all night is annoying — but colder air sits in the bowl. Usually the coldest temperatures in a mountain range are recorded in bowls, windchill factor aside.

5.9 Air moisture

Air accomodates water vapour (moisture) in it. The *relative humidity* of an air mass is the amount of moisture expressed as a percentage of the maximum amount of moisture that this mass of air can contain at the same temperature. The relative humidity is, of course between 0% (perfectly dry air) and 100%. When it is 100%, the air is said to be *saturated*: it can not hold any more water vapour and it starts raining.

The relative humidity depends on the water vapour content of air and on the temperature. When a fixed amount of moist air is cooled, it reaches a temperature at which it becomes saturated, which is called *dew point*. Warm air can accomodate more moisture than cold air, and this is the reason why hot climates feel sticky. The human body cools down by sweating and by the consequent evaporation that cools the surface but, when the air is very humid, there is less evaporation and one feels hot and uncomfortable. The famous Naica cave in the Chihuahua province of Northwestern Mexico, 300 meters underground, has temperatures around 58°C due to the presence of magma below the cave and relative humidity higher than 90%. Because of these extreme conditions, humans cannot survive more than a few minutes in the cave without special equipment and selenite (*i.e.*, gypsum) crystals have grown to the enormous size of 15 meters there because of these special conditions which, fortunately, are not found elsewhere.

In winter the air outdoors is usually drier than in summer because of lower temperatures, and one can see further away on a clear day because the air contains less water vapour than in summer to scatter air (Sec. 7.2). Usually, the views from the same peak are quite different in summer and in winter, not only because winter snow makes distant peaks stick out more, but also because the water molecules in the air scatter light in all directions, blur sharp light sources, and reduce visibility. While winter days can be very clear with high visibility because there is little moisture in the air, haze is associated with the warmer summer air. In fog, the visibility is greatly reduced because light is scattered in all directions and it is impossible to have a sharp image of the source of light rays reaching our eyes or our instruments. Whiteout conditions on snow or on a glacier, where there are no contrast, no shadows, and no perception of depth, create awful situations. In dry air, such as can be found on a clear winter day on a mountain or in a desert, light can propagate through much larger distances without being absorbed and re-emitted or scattered by water molecules. Halos around the moon at night, a sign of less than optimal weather, are also due to moisture which scatters the lunar light.

Generally speaking, however, higher elevation implies less moisture in the atmosphere: mountain air is dry. The largest astronomical observatories around the world are located on high mountains in dry areas. The largest telescopes are located on the high mountains of Chile and on top of Mt. Mauna Kea in Hawaii, where the observing conditions are most favourable. Smaller mountains host less ambitious professional astronomical observatories and amateur observatories (Sec. 7.6).

Infrared radiation and microwaves are absorbed very efficiently by the vibrations of water molecules in the atmosphere. That's why microwave ovens are so good at cooking foods with high water content, and that's why it is very difficult to observe astronomical sources in the infrared and microwave bands from Earth. This is a pity because the cosmic microwave background, which is relic radiation left over from the Big Bang, contains an enormous wealth of information for cosmology [Durrer (2008)]. Microwave and infrared observatories are mostly placed on satellites outside the Earth's atmosphere, although there are some on mountains.

5.10   Weather forecasts

Before going on a committing climb on a high peak, it is essential to obtain a weather forecast as detailed as possible for the area of interest. The weather forecast is not reliable for more than a few days and, usually, not for more than 24 to 48 hours because weather is a chaotic system. This means that the physical laws that rule the fluids in the atmosphere do not allow for much prediction. The physics usually taught in school does not mention this situation. We are used to the idea that physical laws endow us with full predictive power, but this is just one side of the coin. It is true that fluids follow deterministic laws. However, in certain regimes, the solutions of these equations become extremely sensitive to the details of how and where one starts the motion. This means that, by changing initial conditions (the initial configuration and velocity) by only very little, the solutions of these equations (for example the position and velocity of a cold air mass that ruins our climb) are very different. Predictive power is lost and the forecast is bogus.

Chaos occurs often in mechanical systems much simpler than fluids. A double pendulum consisting of a ball attached to a string, which is attached to another ball attached to another string, which is in turn attached to a nail in the ceiling (Fig. 5.8), is very simple, yet chaotic. One swings this double pendulum and stares at it trying to guess whether it will swing right or left: the guess is always wrong and the pendulum is unpredictable. Fluids are much more complicated and even less predictable. Chaos theory has been a conceptual revolution in science in the 1960s and 1970s and is ubiquitous in nature, appearing in physics, chemistry, biology, electronics, and in the financial markets (see [Gleick (2008)] for an excellent popular introduction and [Guckenheimer and Holmes (1983)] for a technical one).

This sad state of affairs with the theory of fluids and air masses can be ameliorated by satellites which track arriving weather fronts instead of trying to calculate their trajectories theoretically, and

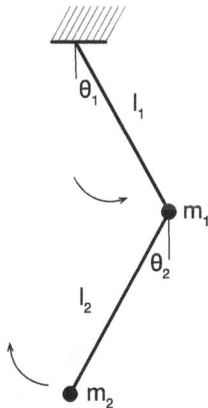

Fig. 5.8   A double pendulum consisting of a simple pendulum with mass m_2 and length l_2 attached to another simple pendulum with mass m_1 and length l_1.

there are repeating patterns, like the winter Azores anticyclone determining long periods of high pressure and stable good weather in the European Alps. But the weather system is intrinsically chaotic and sensitive to small variations in the initial conditions and that's why we cannot obtain forecasts three months in advance to synchronize our climbing trips and our holidays with the good weather.

Chapter 6

Rock climbing

6.1 Introduction

The part of physics that is most relevant for rock climbers is, without doubt, mechanics. It may look redundant to look at physics in sports but it is not so, because human bodies obey physical laws and movement in sports revolves around them. Indeed, the study of physics in many sports is quite interesting [Barrow (2012)], but it is even more interesting in rock climbing. There is much mechanics involved in rock climbing. If the balance of forces between a climber's body and the wall fails, this body falls. When gravity is allowed to apply an unbalanced torque to the climber's body, barndooring (rotation) on one side or the other occurs, which causes the climber to come off the holds. Positioning the centre of gravity between one's feet, flagging legs, and bending one's body sideways are necessary actions to eliminate, or at least diminish, the torque applied by gravity on the body and to avoid this rotation long enough to move to the next hold. Minimizing the force applied by the hands to the holds and the energy expenditure means lasting for an entire rope length on lead instead of burning out in the middle of it.

Anchors keep the climber attached to the wall, but they only withstand a maximum force and usually only if the pull comes from a certain direction, and that's all we have to stop us in case of a fall, so anchor placement must be done carefully by visualizing the force vectors applied to them in case of a fall [Robbins (1971, 1977); Long (1993)]. This does not mean spending extra time in placing gear, in fact often it is crucial to do it quickly because it is tiring or because one has to place protection from a precarious position. One does not worry about this physics when clipping bolts on a sport climb but, in trad climbing[1] and on alpine climbs, anchor placement is an art and a science that takes a long time to learn. The statics of forces at a belay station are crucial to catch a leader's fall and to keep a climbing team alive and they would be the delight of a mechanical engineer. Rappelling from a dubious rappel station should be absolutely avoided but, alas, who has never been in that situation when absolutely nothing else around is available to rappel from and darkness or a storm are approaching as if in fast forward? In this case, minimizing the forces on the dubious anchors is vital and the way to do it is part of the bag of tricks of any climber.

Friction plays a major role, in fact it is the essence of climbing. The rubber sole of rock shoes is prized for maximizing friction against the rock and climbing chalk eliminates sweat from the fingers to increase friction (Fig. 6.1). Too little friction and climbing becomes impossible. Pieces of protection which work well in dry conditions become marginal or useless in wet cracks. When rain hits the wall, only easy climbing and aid climbing remain as options, and protection based on friction should not be relied upon.

Too much friction is an issue, too. When climbing a long pitch, if anchors are not extended with long slings, rope drag becomes a very real problem for the leader higher up. Friction is the key for a safe belay, so that a smaller climber can belay a larger one safely. Again, friction is the reason why knots stay tight (but water knots on slick webbing [Robbins (1971, 1977); Long (1993)] must be checked

---

[1]Traditional or "trad" climbing rejects the use of fixed bolts and removable pitons. Trad climbers jam metal wedges or cam devices in the cracks of the rock. These pieces of protection are removed by the second climber when they climb up following the leader (*route cleaning*). The rock is not permanently scarred by pitons which get inserted and removed by every climbing party, which widens the crack every time, and no piece of gear is left behind permanently to alter the natural appearance of the rock face. This is a non-invasive and the most ethical climbing style but, requiring the leader to place all the pieces of protection, it is also very demanding and is part of alpine climbing.

(a)   (b)

Fig. 6.1   Friction is essential (a) for feet and (b) for hands.

often) and it is the reason why rappel devices allow us to descend a rope safely for a long distance without having to downclimb. Friction can be bad also if it occurs between ropes, or between a rope and a sling. Friction localized in a narrow spot due to rubbing raises the temperature so much that a rope or a sling under tension can easily melt, with horrific consequences on a rappel or when lowering off. This is all physics: while body positioning becomes intuitive and subconscious, there is always much conscious reasoning involved with placing anchors.

6.2   Biomechanics

Apart from psychology, strength, power, and endurance, moving up a steep wall is all about forces and torques, that is, about mechanics. Body positioning is crucial when climbing difficult moves: the forces applied by the climber's weight, the pull by a climber's hand or the push by a toe and the reaction of the rock make the climber go up or fall. Most of the times, this understanding is intuitive and not reasoned as it would be in an engineering course, but the underlying physics is still there.

When beginning to climb, we are told to stand upright on footholds and to push with our legs rather than pulling with our arms because legs are much stronger than arms. Then we progress and we move on to more difficult terrain where the holds are few and far apart, and then these become smaller and sloping. And then we meet cracks, which require a different climbing technique altogether, which involves jamming fingers, fists, and feet.

Easy climbs offer horizontal ledges for feet and hands and climbing is like ascending a ladder, albeit with small rungs. Then, side pulls are discovered: a vertical hold can be used as a side pull but Newton's third law, the law of action and reaction, must be kept in mind. I can pull on this vertical little crack on my right, but the crack applies a force on my hand that is opposite in direction, and equal in intensity, to my pull. This reaction would pull my body sideways and make me lose balance, but fortunately I know Newton's third law and I am prepared. When I pull sideways on the vertical crack, I also position my right foot way out to the right to redistribute the force applied by the crack on me on that decent vertical little ridge sticking out of the wall to my right, below the crack (Fig. 6.2).

Fig. 6.2   The force applied to the wall by the climber generates a reaction applied by the wall on the climber, which must be counteracted using a foot in order to remain in balance.

If I move fast, pulling on the crack, pushing on my right foot, and keeping some tension in my right leg, I can stand up and reach the next good hold far above my head before I start rotating and falling. Thank you, Sir Isaac, for explaining the third law to us.

Another thing that we were told as beginners was to keep our centre of gravity between our feet. Positioning one's centre of gravity between the feet is the way we normally stand and walk on horizontal ground. When the centre of gravity, projected vertically onto the ground, falls outside the space between one's feet, we lose balance and we fall (Fig. 6.3).

If the body leans sideways, we move one foot outward to make this vertical projection of the centre of gravity fall again between our feet, or else we have to lean against a wall or other support. When climbing a snow slope, beginners tend to lean into the slope, motivated by a false sense of security. In this position, the centre of gravity projects vertically outside of one's feet and one has to lean against an ice axe or other support to stay in balance and cancel a component of the weight force. Indeed, it is much better to stand upright on one's feet than to lean into the slope. When the snow slope becomes an ice slope, this position becomes even more important. A climber moving up the slope takes the weight away from the support momentarily when stepping up and that's when a slip occurs unless one is vigilant and stands vertically above one's feet as much as possible.

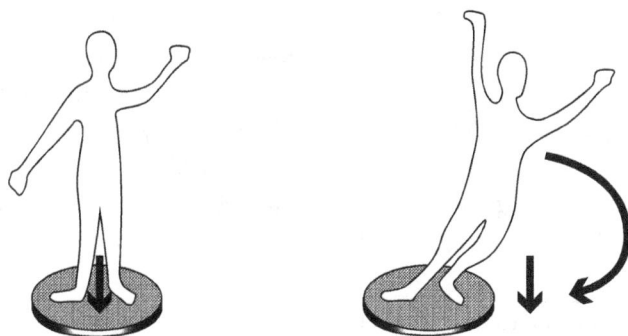

Fig. 6.3 If the vertical projection of the baricentre onto the ground falls outside a narrow area between our feet, we fall.

Fig. 6.4   Lieback, or Dülfer climbing technique.

Back to rock climbing: sometimes one finds a vertical or oblique flake that requires a lieback. The lieback or Dülfer[2] technique requires one to grab a good hold (crack or flake) with both hands and push hard with one or both feet away from it, as if one wanted to rip that hold from the wall, and then one must walk the hands along the flake to move upwards (Fig. 6.4).

A component of the reaction force applied by the rock on the body is upwards and allows one to progress. This technique is very tiring and it is preferable to climb vertical cracks by jamming hands, fists, fingers, and feet instead, but liebacking is useful and sometimes compulsory. In a lieback it is very hard to insert anchors for protection into a crack in the rock because one's body is basically pushed as far away as possible from the crack. In addition, one should make sure to have enough energy to finish a lieback section of a climb because liebacking does not allow rests, it's all or nothing. Deaths

---

[2]From the name of Hans Dülfer (1892–1915), a famous German mountaineer who made some fifty first ascents in the Kaisergebirge of Austria and in the Italian Dolomites before dying prematurely in the Arras front in World War I. He invented the Dülfer climbing technique and a rappelling technique.

have resulted from going a long way on a lieback section of a route without placing protection and becoming tired before finishing it. My friend Antonio witnessed one on a route next to the one he was climbing, and he passed on the wisdom with the crude description of long stripes of blood to our circle of friends.

Another trick that is learned quickly consists of hanging the body's weight from extended arms and not from bent arms (Fig. 6.5). The reason is that, if one hangs from an extended arm, the load is supported by the skeleton and not by the muscles, as is instead the case when one hangs from a bent arm. Lactic acid builds up and muscles tire quickly, but our skeleton does not.

Back to keeping the centre of gravity between one's feet. This is rarely possible in technical climbing and one must apply forces to keep the body from rotating. Sometimes the shift in the centre of gravity is subtle: one stands vertically with the centre of gravity between one's feet, but looking up to search for handholds, one tilts the head backward, which is enough to make the centre of gravity move outside and to lose balance if the pull can not be counteracted. This pull is felt much more when carrying a backpack and wearing a helmet on alpine climbs.

Climbing overhangs requires special techniques. If one climbs with the body facing the wall, it is difficult to reach the holds above and

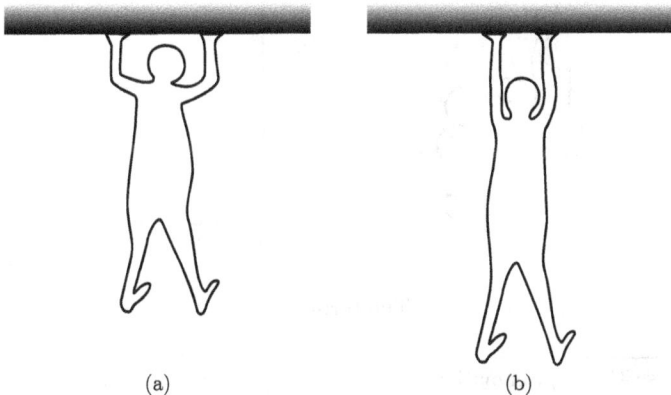

(a) (b)

Fig. 6.5 (a) The weight of the climber is supported by muscular effort. (b) The weight is mainly supported by the skeleton.

gravity pulls down very effectively in the meantime. On overhangs, climbers climb on their side, lock-off on one arm, roll their bodies, twist their pelvis, and reach up with the other arm [Goddard and Neumann (1993)]. This lateral position gives more reach, one climbs with an arm and a leg aligned along a diagonal, and a climber can more easily keep the arms straight.

Now let us consider *torques*, in addition to forces. If a vector force $\vec{F}$ is applied to a rigid body[3] at a point of position $\vec{x}$, measured with respect to a fixed origin O of the coordinates, the torque is the vector

$$\vec{N} = \vec{x} \times \vec{F},$$

where $\times$ denotes the vector product [Fowles and Cassiday (2005)]. The vector product is defined this way: imagine of transporting the vector $\vec{F}$ to the origin O keeping it parallel to itself; now the tails of both vectors $\vec{x}$ and $\vec{F}$ are at the origin O (Fig. 6.6).

$\vec{N}$ points perpendicular to both vectors $\vec{x}$ and $\vec{F}$. The two vectors $\vec{x}$ and $\vec{F}$ define a plane. In this plane, imagine of bringing the vector $\vec{x}$ over the vector $\vec{F}$ by keeping its tail fixed and rotating it through an angle less than $180°$. Curl your right hand and let the fingers follow the motion of the vector $\vec{x}$, then the thumb stretched out gives the

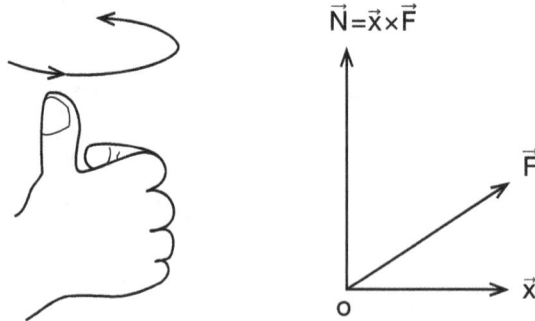

Fig. 6.6   The torque $\vec{N} = \vec{x} \times \vec{F}$.

---

[3]Let us use the approximation of a rigid body for simplicity, as done in most mechanics courses. The rigid body is not a good approximation if abdominal and gluteal muscles are not very strong, but in that case the modelling becomes really complicated and the climbing becomes ineffective.

Fig. 6.7 Barndooring.

direction of $\vec{N} = \vec{x} \times \vec{F}$ (Fig. 6.6). In practice, a torque will cause a rotation of the body in space about an axis passing through the origin O (Fig. 6.7).

The magnitude of the torque is $|\vec{N}| = |\vec{F}||\vec{x}| \sin \theta$, where θ is the angle between the vectors $\vec{x}$ and $\vec{F}$ (incidentally, this magnitude is equal to the area of the parallelogram spanned by $\vec{x}$ and $\vec{F}$). In other words, the magnitude of the torque depends on both the magnitude of the force applied and on the geometry. If the angle θ is fixed, the torque magnitude $|\vec{N}|$ is larger if the magnitude of the force applied is larger, or if the distance $|\vec{x}|$ from the rotation axis is larger. However, the angle θ is also important. Given equal $|\vec{F}|$ and $|\vec{x}|$, the torque will be maximum if the angle θ is 90° (corresponding to $\sin \theta = 1$), meaning that the force is perpendicular to the distance from the rotation axis $\vec{x}$. The torque will be zero if the force $\vec{F}$ is parallel to the position vector $\vec{x}$, *i.e.*, the oriented distance from the rotation axis, corresponding to $\theta = 0°$ and to $\sin \theta = 0$. As an example of a torque, consider the situation when we open or close a door (Fig. 6.8).

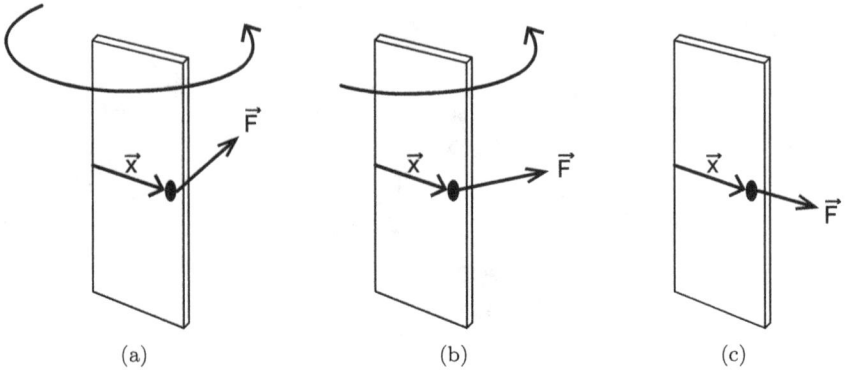

Fig. 6.8   Opening a door. (a) Maximum torque and maximum effect for $\theta = 90°$, or $\vec{F} \perp \vec{x}$. (b) Torque somewhere between zero and maximum, some effect for $0° < \theta < 90°$. (c) Zero torque and no effect for $\theta = 0°$, or $\vec{F} \parallel \vec{x}$.

To reduce effort, we want to minimize the magnitude $|\vec{F}|$ of the force that we apply to open the door. Therefore, we place the handle far away from the vertical rotation axis passing through the hinges, which maximizes $|\vec{x}|$. Then, we apply the force $\vec{F}$ so that its direction is perpendicular to the plane of the door, $\theta = 90°$, which maximizes $\sin \theta$. If we try to open or close a door by pulling it horizontally in the plane of the door, with the force $\vec{F}$ parallel to the position vector $\vec{x}$ of its point of application (the door handle), we have $\theta = 0°$ and $\sin \theta = 0$ and the door does not rotate at all (Fig. 6.8). So, by maximizing $|\vec{x}|$ and $\sin \theta$ we minimize the magnitude $|\vec{F}|$ of the force needed for the minimum torque that, applied to the door, closes or opens it comfortably.

In climbing, it often happens that handholds and footholds are not placed like on a man-made ladder, but they are all off to one side of where we are standing. Holding a handhold, say, on the far left with the left hand, and placing the left foot on a foothold on the far left, with no other holds for the right hand and foot, creates a torque applied on the body by the weight force. This torque makes the body rotate and takes the foot and hand away from small or sloping holds. To eliminate the torque, one flags the right leg to the left, which takes part of the mass away from the right side of the body and redistributes it to the left. The distance $|\vec{x}|$ of some masses in the body from the rotation axis then becomes zero or small, so the strength $|\vec{N}|$ of the torque vanishes or decreases significantly. Sometimes one can move even more body mass to the left, by placing the right foot instead of the left one on that foothold to the far left, crossing legs in doing so. In this position, the outside edge of the right shoe touches the wall. Normally there isn't much room on a vertical wall to do this leg flagging and crossing, but there is plenty of space on overhangs, where knees don't bang against the rock when they move around. Flagging a leg to counterbalance the weight and to avoid barndooring requires an understanding of torques, even if it is intuitive and not reasoned (Fig. 6.9).

As in every sport, one coaches the body to learn and remember these movements automatically without thinking about them [Goddard and Neumann (1993)]. However, it takes some conscious thinking when first learning those moves and positions. The pull of gravity, the forces applied, and the torques due to the peculiar distribution of weights around the pivot points are very real.

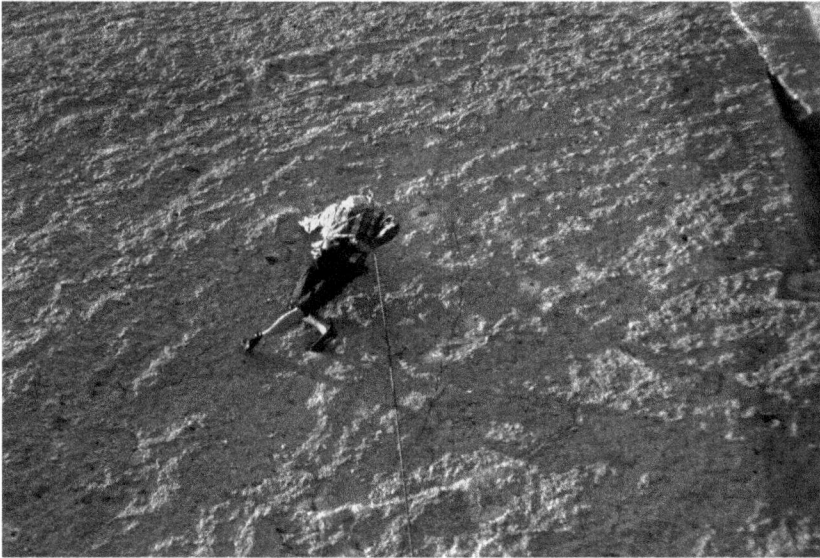

Fig. 6.9 A climber flagging his right leg to avoid barndooring on Heinous Cling, 5.12 (Smith Rock, Oregon). The rotation axis goes through the left hand and left foot.

6.3 Fall factor

A rock climber is climbing a wall on lead when he takes a fall (Fig. 6.10).

The shock of the fall must be absorbed by the elasticity of the dynamic rope, possibly a slight slip of the rope in the belay device of

Fig. 6.10 A fall with a small fall factor.

the belayer, and the system of protections set up by the climber. The UIAA (Union Internationale des Associations d'Alpinisme or International Climbing and Mountaineering Federation) came up with a standard for climbing ropes: they must be able to take a 12 kN fall. The rationale is that this is the maximum force that a human body can withstand without dramatic injury. The other components of the safety system (harness, fixed anchors, and carabiners) are designed to take a larger force. Essentially, the stretch in the rope must absorb the impact. Alpine climbs usually involve long runouts. The thought of a screamer (long fall) on one of these runouts is a nightmare and the danger is real. However, it's not only these long falls that are dangerous, and this fact comes as a surprise to many climbers. The load on the anchors at the belay station, on the pieces of protection, and the shock on the falling climber's body when the fall is arrested can be quite large even on a short fall, if there is little rope stretch. It's no joking matter and the fall factor was designed to quantify the danger and to include these situations.

The fall factor is the length of the fall divided by the length of the rope paid out by the belayer (the length of rope going from the belay device to the leader),

$$\text{fall factor } F = \frac{\text{length of fall}}{\text{length of rope out}}.$$

The length of the fall is twice the distance between the point where the leader falls and the last protection connecting the rope to the wall, assuming that this piece of protection holds. It is intuitive that, the longer the length of rope paid out, the more stretch there will be in the rope, so the denominator of the fraction defining the fall factor must be important.

The worst case scenario is when the leader has not placed any protection and falls directly on the belay station, which corresponds to the maximum value of the fall factor, 2 (the length of the fall cannot exceed twice the distance from the anchor). Since the rope must absorb the shock and the belay system is designed around this property, it is clear that a short fall with fall factor 2 is bad because there is almost no rope out to absorb the shock of the fall — the situation is very similar to a fall on a static rope that doesn't stretch and transmits the full force to the climber's body and to the anchor. The force developed on an $F = 2$ fall of a few meters is comparable to that developed on a 20–25 meters fall.

Another very dangerous situation occurs when a climber ties into a belay station with a sling and then climbs above the anchor to scout the next move, to take a picture, or for whatever silly reason, and then falls on the anchor. The fall factor is 2. There are static slings sold in climbing stores which are considered to be okay if used as runners for the carabiners connecting the dynamic rope to the pieces of protection placed by the leader, because they are part of a larger and complex dynamic system with sufficient elasticity built in, but they are not okay for taking a hard fall with $F = 2$. Fatal accidents have happened in this kind of situation at belay stations.

When the leader progresses up the pitch and places solid pieces of protection which stop a fall without pulling out, the fall factor is always less than 2 and the longer the length of rope paid out by the belayer, the more it will stretch. It is clear that the more pieces of protection are placed, the safer the fall of a leader is but, in practice, we don't always find a crack that takes a piece, or we don't have a piece of the right size, or we cannot carry an infinite rack. What is more, speed is often safety in the mountains [Twight (1999); Simpson (1989)] and the extra time needed to place extra protection is sometimes detrimental. In many mountain areas, one wants to summit and get down before the afternoon thunderstorm comes, not to mention the extra rope drag that becomes significant when many pieces of protection are placed on a long pitch. Good judgment and plenty of cracks for stoppers and friends (anchors) are essential for protection, failing which climbing proficiency and luck come next.

There is another situation in which a fall factor 2 or larger occurs. On the via ferratas common in Europe, climbers clip into a fixed steel cable with two carabiners connected with two slings of fixed length to their harness. The steel cable is anchored to the rock (Fig. 6.11).

Fig. 6.11 A via ferrata.

On vertical stretches, a via ferrata climber who falls will pass the first piece of iron below which attaches the steel cable to the wall and will then continue on the way down for the entire length of his or her slings, followed by the two carabiners, which eventually smash onto said piece of iron. The fall factor exceeds 2 and, to make things worse, the carabiners do not work lengthwise, which is what they are designed for, but they are oriented sideways and they are a lot weaker in this configuration. The shock force developed in such a fall is more than enough to break the carabiners or the climber's body and this is a recipe for disaster. To mitigate the severity of the impact, a device designed to dissipate the energy of the fall is essential. Until not long ago, this device used to be a rounded plate (*dissipatore*) in which a piece of 10 mm rope passes through several holes and slides with much friction upon an impact. Nowadays the system of choice to dissipate the energy of the fall is a piece of webbing sewn rather loosely onto itself several times. A fall rips these seams and extends the sling to absorb the energy of the impact. But, in any case, a vertical fall with high fall factor on a via ferrata is always a dangerous scenario and must be avoided at all costs.

6.4   Never rappel from a tight sling

During a hot summer day many years ago, I was climbing a rock route three or four pitches long (I'm not sure because I never finished it) in the Adige River valley of Italy. We were four climbers, split into two parties with a single rope each. This detail is important because you needed two 60 meter ropes to rappel off that route. As it often happens there on a hot summer day, we saw a thunderstorm coming and we all prepared for a hasty retreat. I mean, all of us except for my climbing partner who, having already climbed a few meters up the third pitch, which was only 25 meters long, decided to climb it all the way instead of lowering off to rappel down before the thunderstorm hit. A perch on a high wall is not the best place to watch lightning bolts from. But it was his turn to lead and he wouldn't miss it for any reason. We could not convince him that we had to head down as soon as possible. No amount of clever reasoning, followed by less polite argumentation, calls to self-preservation instincts, and then profanities, worked and he kept going. Things didn't look good. He made it to the next belay station and then he rappelled down, the process taking what felt like an eternity. Only later we learned that, not satisfied about taking a chance with the thunderstorm, he had taken a bigger chance rappelling from a tiny sling tied very tight between the two pitons forming the belay station. The rationale for this choice was simple and perfectly logical: he wanted to leave behind the smallest, shortest, and therefore cheapest, sling that he had and nothing more. Somehow he survived that rappel.

By the time my rope partner was down to the beginning of the third pitch, where we were unhappily waiting for him (and, selfishly, for the second rope needed to rappel), the thunderstorm was sweeping our wall. First we were bombarded by the most hail that I have ever seen come down in such a short time. I recall finding some consolation in the fact that I was wearing a helmet. Then torrential rain followed. There was no point in trying to fend off the torrent or even thinking of keeping something relatively dry. We were totally exposed and it was like being plunged under water. In the high wind we had a hard time retrieving the rappel rope from the higher station. Fortunately we didn't have any lightning nearby and after fifteen minutes or so the rain eased off, the wind dropped, and we had our first rappel set up. We didn't know whether the storm that got us wet, cold, and scared was circling around or leaving, so we postponed the mandatory insults. With water rushing down the wall, slippery rock, a ledge below turned into a lagoon, and cold hands and everything else, we finally rappelled to the ground where a marvel was waiting for us. A giant pile of hail similar to a talus cone adorned the base of the wall along its entire width. All the hail that had bounced off the wall was now piled up at its base in steep cones. Incidentally, because hail is a granular material it is appropriate to talk about its angle of repose (Sec. 2.3), which is fairly large. In glaciology, snow mixed with graupel (very similar to hail, but technically distinct from it for snow and ice scientists) is known to have a large angle of repose of $45°$ [Abe (2004)]. The hail pile was very deep and it was a strange sensation to land in it and wade through.

When we later learned about the sling left on the highest station, which was now missing from my partner's harness, we were horrified. What is wrong with rappelling on a tight sling is a classic exercise in a Physics 101 course.

Think of a sling anchored at two points, for example two pitons, aligned along a horizontal line (Fig. 6.12).

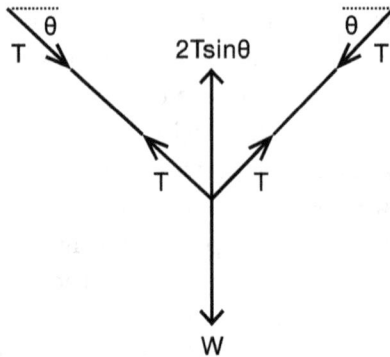

Fig. 6.12   A sling connecting two anchors. The tension is the same in both sides of the V, and the vertical force balancing the weight W is $2T\sin\theta$.

The sling supports the weight W of a climber attached to its midpoint and it sags, forming a V. Let θ be the angle that the sling makes with the horizontal at each piton. By symmetry, this angle is the same at these two points and the tension in the sling has equal magnitude T in each half of the sling. The tension in each half of the sling is a vector $\vec{T}$ which originates in the midpoint and points along the sling toward the piton located on that side of the sling. Each one of these two vectors pointing toward a piton has the same magnitude T. The weight of the climber hanging from the midpoint is a vector pointing vertically downward, it is applied in the midpoint of the sling, and has intensity $W = mg$, where m is the mass of the climber (my friend was a tall guy of considerable m) and g is the acceleration of gravity. In static equilibrium, the vertical components of the tensions on each side of the sling's midpoint add up to balance the weight W. The horizontal components are balanced by the reaction of the anchor points. So, let us decompose the tension T on each side into horizontal and vertical components. Trigonometry gives the vertical component on each side as $T\sin\theta$. Since there are two sides, the balance of forces in the vertical direction is simply $2T\sin\theta = W$, which gives us the dreaded formula for the tension on each side

$$T = \frac{W}{2\sin\theta}.$$

You see immediately what's wrong with this asinine way of rappelling: when the sling is tight the angle θ is small, close to zero degrees. When an angle is small, its sine is also small (calculus tells us that $\sin\theta \simeq \theta$ for $|\theta| \ll 1$) and this sine is in the denominator so, as θ tends to zero, the tension T tends to infinity! Infinity is an idealization of course, but the point is that the tension can reach very high values, sufficient for the sling to snap. We can insert numerical values in our formula for illustration, say

$$T = \frac{(90\,\text{kg})\,(9.8\,\text{m/s}^2)}{2\sin(3°)} = 8400\,\text{N}.$$

By comparison, the weight of the climber is

$$W = mg = (90\,\text{kg})\,(9.8\,\text{m/s}^2) = 880\,\text{N},$$

so the setup magnifies the load on the sling by a factor 9.5. The tension in the sling, and the force applied on the anchor points, are much larger than the weight of the climber. True, the sling may stretch and sag a little more and the angle θ tends to increase. Or the sling may snap, and tight slings do snap. A dynamic rope with a 10 mm diameter is designed to hold 12 kN, but a sling of 5–6 mm diameter holds much less.

The second point to make is that, even if the sling is very strong, the tension T is applied to each anchor point and a small angle θ multiplies the force on each anchor. Weak anchors fail, and they may not be equally strong with respect to every direction of pull. The force applied to an anchor in case of a fall should never be multiplied by an improper setup.

The correct way to set up a rappel or belay station is to use a long sling so that, when the load is applied in the middle of it, the angle θ on each side remains large (Fig. 6.13). If we use the numbers above and we change θ from 3° to 45°, we obtain $T = 620\,\text{N} = 0.7W$ and the tension on each side of the sling is now less than the load applied to its midpoint. The angle θ must be as large as possible. In other words, the angle made by the two halves of the slings at the sling's midpoint (the bottom of the V) must be as small as possible.

Devil-may-care attitude comes with a high price in climbing and my friend, who clearly was not a physicist, is lucky to be alive. Never rappel from a tight sling.

6.5   Beware of the American triangle

Belay and rappel anchors must be sufficiently strong to stand all the pulls that climbers apply to them. Usually one needs two or more anchor points for a belay or rappel station, unless a single point is bombproof, such as a bigger tree growing on a wall. The correct way to set up a station with two anchor points is by connecting them in such a way that both points stand equal forces, that is, by equalizing the anchors. One should also avoid applying unnecessary forces to the anchor point, which could fail if the applied force exceeds a threshold. The American triangle is an old setup which is hopefully disappearing fast from climbing practices, but was once common, also in Europe. This setup multiplies the forces applied to the anchors and must be avoided.

Consider two anchor points: the correct way to use them is by applying two slings of equal length to each point, or to attach a long sling and equalize the load on the anchors [Robbins (1971, 1977); Long (1993)] (Fig. 6.13).

The dangerous American triangle, intead, uses a single sling in a V-shape passing through both anchor points, with the climber attached to the bottom of the V (Fig. 6.14).

Let us analyze the forces in the American triangle. Consider the tension in the sling: each anchor point is subject to the tension in the arms of the V, pointing along the sling, *plus* an additional horizontal tension due to the fact that the sling passes through the anchor point to connect to the other arm of the V (Fig. 6.15). This extra tension is absent in a correct setup in which each end of a long sling is connected to each bolt (Fig. 6.13).

Let us use the example of the previous section. A climber with a mass $m = 90$ kg is attached to the bottom of the V, and the angle there is $\alpha = 45°$, which means that $\theta = 90° - \alpha = 45°$. As already seen, the tension applied to each anchor and pointing along the sling

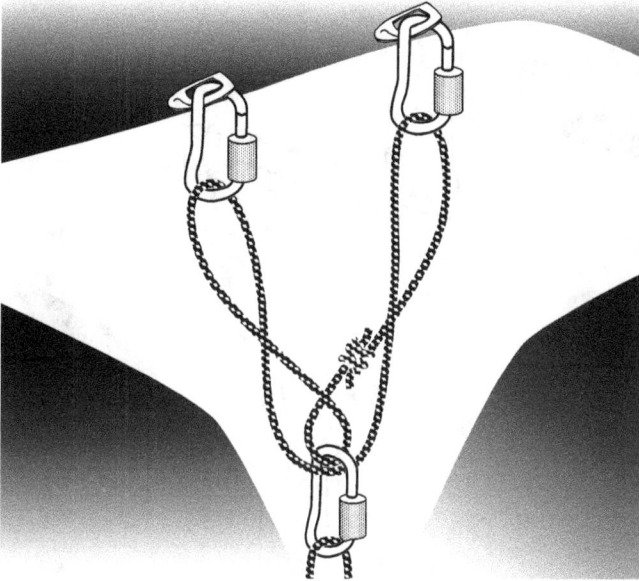

Fig. 6.13   A station built with two equalized anchor points.

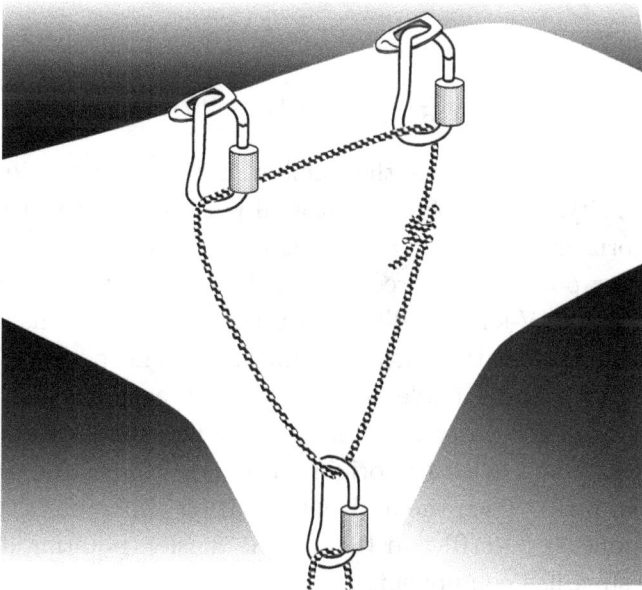

Fig. 6.14   Never use this American triangle for a belay or a rappel station.

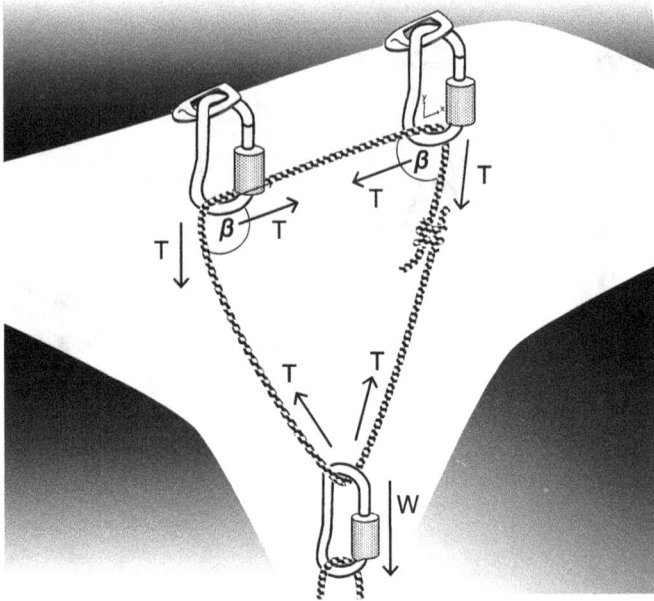

Fig. 6.15   Forces in the American triangle. Never use it.

has intensity

$$T = \frac{mg}{2\sin\theta} = 620\,\text{N} = 0.7W,$$

where W is the weight of the climber. But now a tension of the same intensity, and pointing horizontally, is applied to each anchor by the horizontal segment of the sling. The total force applied on each anchor points in a direction which is somewhere between the direction of the V-shaped sling and the horizontal and has a larger intensity than with the correct setup. The angle between the two sides of the triangle made by the sling at each anchor point is $\beta = (180° - 2 \cdot 45°)/2 = 45°$. Let (x, y) be Cartesian coordinates in the vertical plane, with the origin located at the rightmost anchor point (Fig. 6.15). Then the horizontal tension applied to this anchor has components $(-T, 0)$ and the oblique tension pointing along the V of the sling has components

$$(-T\cos\beta, -T\sin\beta) = (-440\,\text{N}, -440\,\text{N}) = (-0.7W, -0.7W).$$

The total force applied to each anchor point has components

$$(F_x, F_y) = (-T - T\cos\beta, -T\sin\beta) = (-950\,\text{N}, -440\,\text{N})$$
$$= (-1.7T, -0.7T)$$

and an intensity

$$F = \sqrt{F_x^2 + F_y^2} = 1100\,\text{N} = 1.8\,T = 1.3\,W.$$

The force on each anchor is multiplied unnecessarily. If the anchor points are not bolts or solid pitons but are instead stoppers or cam devices jammed in cracks, which do not hold equally in all directions, the anchors can fail. Forces on the anchor points should never be multiplied unnecessarily. Avoid the American triangle.

6.6 The yin and yang of friction

Friction is good for you; friction is bad for you, it's the usual yin and yang. Think of the friction between rock and the soles of our rock shoes. We want just the right amount of load on the footholds: when stepping on a polished sloper, being timid and loading too little weight means not enough friction between the sticky rubber sole of your climbing shoe and the rock, and a slip is guaranteed. Being cocky and carelessly loading all the weight on one foothold means again slipping if the hold is too smooth to provide enough friction. What is needed is just the right load and the right amount of friction. Remember that the friction force has magnitude $F = \mu_s N$ proportional to the normal force N. Slab climbing (Fig. 6.16) is a game of friction and nerves. These slabs are usually much less inclined than vertical and have practically no holds, save for a few tiny warts spaced far apart. Hands cannot do much more than pushing lightly against the rock to keep in balance and looking for microscopic features to hook one's fingernails on, while feet do all the work to push the body up. Because there are no footholds either, the game is all friction. It is unnerving to move in order to step up from a position where one is barely standing on both feet. Once a move is made, repeat many times. The happy thought that a fall on a slab is not

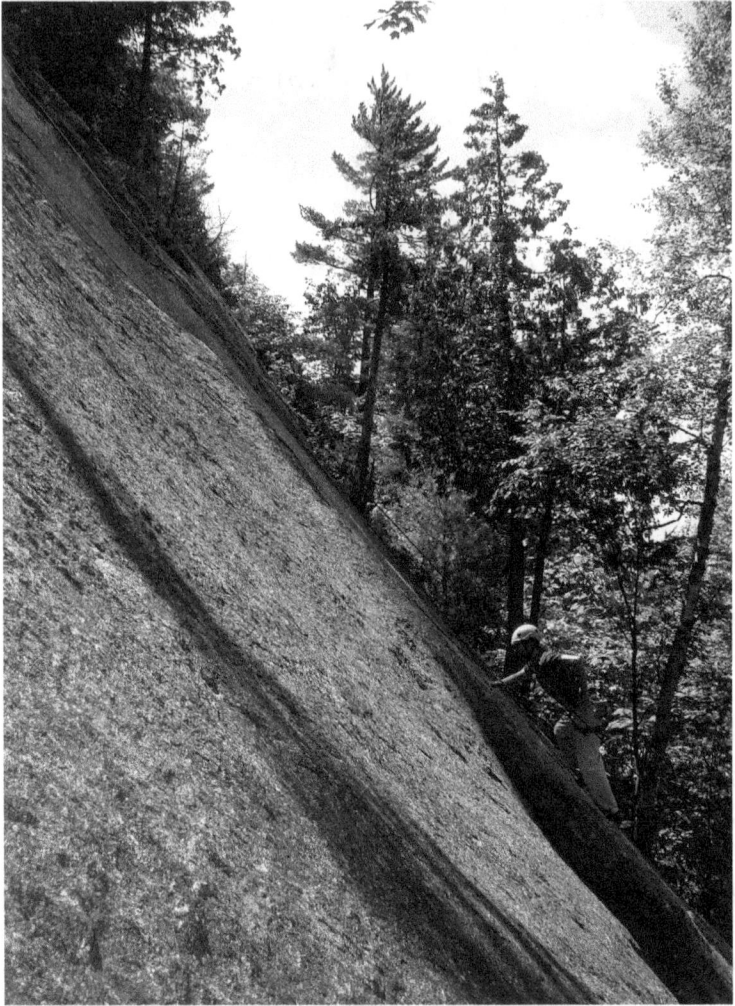

Fig. 6.16   Slab climbing practice (Mont Pinacle, Québec, Canada).

nearly as bad as a fall on a vertical wall makes slabs traditionally poorly protected, which compounds the problem that insane amounts of sheer faith in the friction of our soles is required to climb them. As a result, slab climbing is very specialized and slab climbers are very skilled and balanced maniacs with huge calf muscles and steel nerves.

Think of the friction between the climbing rope and the belay device,[4] which is how a leader's fall is stopped. Not enough friction means that a fall is not stopped, or it is not stopped soon enough. However, if a fall is held by a snow anchor, much weaker than a solid piton or bolt in a rock wall, we want a very dynamic belay with little friction in order not to overload this anchor.

A half-hitch knot on a carabiner, without a belay device, may provide sufficient friction for a belay, but not enough to stop it immediately and it may put a large load on the anchor, and that's how it was done in the old days. On alpine climbs on easier rock or on snow, the half-hitch knot on a wide carabiner is still a quick way to set up a belay with minimal gear. The half-hitch knot weakens the strength of a rope in case of a fall and puts kinks in the rope, but in some situations there are other priorities, such as speed. Friction is what makes up a belay.

Other times friction is not so welcome. Near the end of a long pitch of trad climbing, the friction of the rope against the rock becomes a curse, especially for a climber like this author who chickens out and deploys all his rack in every possible crack of the route to ensure real or illusionary protection. Using long runners to connect the rope to the pieces of protection and placing gear wisely and sufficiently spaced minimizes this unwanted friction.

Quite the opposite, the pieces of protection that the trad climber sticks in the available cracks work because of the friction between these pieces of metal and the rock and, in this case, the more friction the better. Stoppers (metal wedges) are jammed against the walls of the cracks. The cams of camming devices (*friends*) are engineered to maximize their friction against the rock and they hold much less if a crack is lubricated by seeping water or rainwater. In fact, even the sweat from our hands is enough to send us down on small or medium handholds, and climbing chalk is designed to eliminate the sweat and increase friction.

Then there is friction between ropes, which may be a good thing if we use a Prussik sling tied onto the main rope as an emergency

---

[4]Commonly called Air Traffic Controller (ATC).

brake during a rappel, or if we use two Prussik slings to self-rescue and climb up the rope out of a crevasse after falling in. A friend of mine used two Prussik knots to climb twenty-five meters up the rope on an overhang after falling in a long pendulum on a traverse on loose Canadian Rockies limestone. The friction between the Prussik knots in the slings and the climbing rope avoided a tricky rescue.

But rope-to-rope friction is not always good. Ropes under tension that rub against each other melt very easily. I heard about a beginner climber who was top-roping a sport route with his rope running through a sling left on the belay station instead of running through the mandatory metal carabiner. Under tension, the sling melted, sending him to the hospital in critical conditions. When rappelling, instead, it is normal to pass the rappel rope through a sling left behind on the rappel station, but the rope must not move relatively to the sling. Because of the load, a small movement is accompanied by high friction which can melt both sling and rope. I remember a sling found on a station where somebody had rappelled before me: it was melted halfway through, a reminder that under load even a small slip of the rope is a bad thing. If a rappel station is used a lot, a locking carabiner or rappel rings must be left in place for use by the community (Fig. 6.17).

Then there is friction between the rope and the rappel device, which heats up both, but is particularly dangerous for the former. Friction is particularly high in free-hanging rappels when the entire weight of the body is placed on the rappel rope (Fig. 6.18).

A fast rappel must be avoided because it generates way too much heat which damages the rope, unless we are rappelling in torrential rain which dissipates the heat quickly (I remember a few such occasions), or unless we are canyoning and then the ropes are always wet (Fig. 1.2). The Navy Seals are allowed to rappel very fast, but only because they have bigger concerns in their missions.

From another point of view, think of the enormous friction of a glacier against its valley walls, its bed, and the rock it scours during its motion. Tens of thousands of years later we see the effects of this friction as U-shaped glacial valleys (Sec. 4.6), round polished rocks, and as deep scars left on softer rocks. Indeed, the U-shape of glacial

Fig. 6.17 A rappel station. Note the rappel ring and the small angle formed by the V of the chains.

valleys is modelled theoretically using a variational method which maximizes the friction against the valley walls (Sec. 4.6).

What is friction? In physics there are various kinds of friction but they all have something in common: friction is a force that always opposes motion and, if there is motion, it converts mechanical energy into heat. This last part is well known because we rub our hands fast and hard to warm them up when they are cold. Similarly, when we take a rappel device off the rope at the end of a long rappel, especially a free-hanging one, we find it very warm.

Static friction occurs when two objects are in contact but do not move: if a force acts on one of them to make it move with respect to

Fig. 6.18   A properly executed free-hanging rappel.

the other, an equal and opposite friction force arises which impedes this motion. An example is a block of some material placed on a plane inclined with the horizontal, as seen in Sec. 2.3. If the friction is sufficiently large, the block does not slide along the slope. In general, the formula for the intensity of the static friction force is

$$F = \mu_s N,$$

where N is the force normal (*i.e.*, perpendicular) to the plane and μ_s is a pure number (*i.e.*, without units or dimensions) called the coefficient of static friction.

Dynamic friction occurs when there is motion between two objects, for example between a block and a horizontal plane along which it is pushed. Again, the friction force opposes the motion so, if the block is moving horizontally along this plane, the friction force is a vector which is horizontal and points in the direction opposite to the motion. Its intensity is given by the so-called Coulomb law of friction[5]

$$F = \mu_k N,$$

where N is again the normal force (vertical) and μ_k is the coefficient of kinetic friction, another pure number. We have already applied this formula to objects sliding down an incline in Sec. 2.3. Now think about two ropes rubbing against each other under tension. If N is large, then F is large and that's why even a small movement of the two ropes against each other under a heavy load N means high friction and a relatively large amount of energy dissipated into heat. The high temperature generated in that spot can melt a rope or a sling. For this reason, a rope, sling, or piece of webbing under tension should never be allowed to move, especially if it rubs agains another rope or against the rock. One must take care not to let the rappel rope hanging from a sling to slide in it when this rope is weighted or unweighted during the rappel. A smooth descent during which the rope is constantly under the same tension without sudden jerks is the best way to rappel (moreover, bouncing on a rope that is running over a rock edge can saw it).

Viscous friction is friction that arises when there is motion, but it requires a very different model. The dynamic friction seen above

---

[5] Another, better known, Coulomb law in electrostatics has nothing to do with this equation for friction.

has no dependence on the velocity, while viscous friction depends on the relative velocity of the objects in motion. At relatively low speeds, the usual model of viscous friction (*e.g.*, Halliday, Resnick, and Walker (2005); Fowles and Cassiday (2005); Goldstein (1980)) assumes that the intensity of the friction force is proportional to the velocity, $F = \alpha v$. At high speeds, the friction force is proportional to the square, or even to higher powers, of the velocity, $F = \beta v^2$. These models of friction are particularly appropriate to describe the friction encountered during the motion of an object in a fluid, and they have been used in the previous chapters.

Chapter 7

Miscellaneous

7.1 Introduction

Many very useful objects are thrown in the junk drawer of a kitchen, not because they are actually junk, but because we do not know to which of the other, very ordered, drawers and cabinets they belong to. Psychologists say that, indeed, the human brain is very good at filing away and classifying memories and concepts and the process of organizing a kitchen or hardware store is not much different. When some object or some concept doesn't fit neatly in a certain classification scheme or set of boxes, we create some sort of junk drawer to which the object or concept is doomed to [Levitin (2014)]. The same happens in this book, and in all books attempting to cover a broad range of topics, which contain a "miscellaneous" area. This last chapter contains sections that don't quite belong to the previous chapters and that can not be forced there without regret. This "junk drawer" of the book could, of course, be much longer, and the book itself could be extended *ad libitum*, but completeness is not our goal here. I have included a few sections on topics that caught my mind on recent climbs and that I did not want to pass on. Some of them are common to most alpine climbs.

Let us begin again with the obvious. In good weather, looking up in the sky we see it bluer and bluer, while looking down on the horizon we see it whiter. We see it red at sunset and sunrise. These facts are so familiar that normally we don't even pause to question why things are that way, but their physical explanation is far from trivial. Microscopic scattering processes involving molecules of atmospheric gases and other particles are at play and their detailed explanation requires quite a bit of physics [Svanberg (1992)].

In bad weather, looking up in the sky we do not see much and we may be better off worrying about getting out of there, especially if we see very dark clouds approaching. This is the case when electricity builds up and an electric storm is arriving. If getting out in time is not possible, reaching the safest possible place under the circumstances should become a priority. Some basic understanding of electromagnetism may help in these situations. If, instead, we are safely at the bottom of the valley inside a dry tent, or perhaps in front of a drink inside a warm hut in the European Alps, we can admire the lightning bolts hitting the mountain tops nearby and enjoy safety and dryness.

Many less spectacular things are encountered again and again in alpine terrain, for example, lichens. Sometimes lichens colour an entire wall orange or green, other times we see only small patches of lichens growing timidly on the rocks, and other times again they stubbornly decorate all the holds of our rock route. And, occasionally, we see fairy rings of mushrooms in the meadows below the alpine. They have something in common with lichens. There is some interesting science, in addition to the obvious biology, involved in the way lichens and fairy rings of mushrooms grow. The list of interesting alpine things could go on and on, but I decided to end our physics trip with a brief mention of facilities for fundamental science actually located in the mountains and requiring the mountains to operate.

7.2   Why is the sky blue?

Assuming that you had blue skies on your mountain trip today, you might have been wondering about the most obvious question of all: why is the sky blue? Why is it instead red at sunset and sunrise? Why is the sky darker at higher and higher elevations, and it is darker looking straight up than looking toward the horizon?

These questions may be obvious, but the answers are not. The common answer to these questions involves *scattering*, a physical process in which photons (quanta of light) impinge on molecules or other particles [Svanberg (1992)]. There are several types of scattering of photons on other particles. They are classified according to their dominant physical characteristics with respect to size of the scattering particle, energy of the incident photon, *etc.* Light from the sun scatters against the molecules and particles of the atmosphere, but not all the wavelengths that compose the solar light are scattered equally. Some are scattered more and some less, according to the circumstances. Electromagnetic waves, including light waves, propagate at the speed of light. The speed of light in the atmosphere is almost identical to the speed of light *in vacuo* $c = 300,000 \, \text{km/s}$. Think of a sinusoidal wave (Fig. 7.1): the distance between two consecutive peaks (or between any two points in phase) is the *wavelength* λ (measured in meters) and the number of cycles (the number of times the wave repeats itself) per second is the *frequency* ν, measured in Hertz $(1 \, \text{Hz} = 1 \, \text{s}^{-1})$ [Halliday, Resnick, and Walker (2005); Main (1993); Pain (2005)].

The frequency is the inverse of the *period* τ of the wave, which is the time occurring between the passing of two consecutive peaks in a point of space, $\nu = \tau^{-1}$. Since the speed of propagation of the wave is space travelled over time taken to travel it, a wavelength

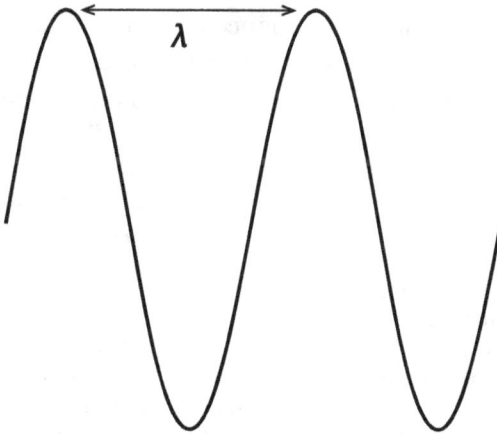

Fig. 7.1   A sinusoidal wave.

λ is travelled at speed c in the time of one period $\tau = \nu^{-1}$ and

$$c = \lambda\nu.$$

This is the fundamental relation between wavelength, frequency, and speed for any sinusoidal electromagnetic wave. Substitute the speed of light with the speed of propagation in a medium, and the same relation is true for any simple wave of whatever nature, sound wave, wave in a string, *etc.* [Halliday, Resnick, and Walker (2005); Main (1993); Pain (2005)].

Blue light, which corresponds to the shortest wavelengths and the highest frequencies of the visible spectrum, is scattered in all directions more intensely than light of longer wavelengths by the molecules of the atmosphere as it travels through it, explaining why clean skies are blue. This phenomenon was (almost) explained by John Tyndall [Tyndall (1868)], the prominent Irish physicist and mountaineer famous for being the most active science popularizer in the mid 1800s and for the first ascent of the Weisshorn in the Valais region of Switzerland. Indeed most, if not all, of his contributions to science, including the discovery of the now infamous greenhouse effect causing global warming, originated from his mountaineering [Tyndall (1872); Reidy (2010)].

The scattering by molecules of the atmosphere, called Rayleigh or elastic[1] scattering, is more pronounced at short, blue wavelengths. In fact, this scattering depends strongly on the wavelength. The intensity I of the scattered light is proportional to the inverse fourth power of the wavelength of the light scattered,

$$I(\lambda) \propto \frac{1}{\lambda^4},$$

---

[1] *Elastic* means that the photons incident on the molecules do not gain or lose energy during the scattering process. If they do, their frequencies and colours are shifted — blueshifted to higher frequency if they gain energy, redshifted to lower frequency if they lose it. This frequency shift characterizes inelastic, or Raman, scattering. Inelastic scattering is due to the fact that the photons excite, or de-excite, vibrations of the target molecules losing or gaining energy in the process.

hence longer wavelengths are scattered little, while short blue wavelengths are scattered much more intensely.

Red light, which has longer wavelength than blue light, is not scattered much by the molecules in the atmosphere through the Rayleigh process. However, it is scattered intensely by large particles in Mie scattering if these are present. This other type of scattering depends only weakly on the wavelength of the light scattered and gives the white glare observed around the sun and the moon when there is fog or humidity in the atmosphere. Here large particle size means comparable to, or larger than, the wavelength λ of the light scattered. Particles larger than the wavelength of red light ($\sim$700 nm = $7 \cdot 10^{-7}$ m) include dust, sand, aerosols, and the particulate matter responsible for much of man-made pollution. This is the reason why polluted areas near cities often produce attractive sunsets and why the red colors of sunset appear only when the sun is lower on the horizon and its rays traverse the bottom part of the atmosphere, where sand, dust, smoke, aerosols, and impurities are concentrated. In addition, at sunset and sunrise the rays from the sun pass through a greater thickness of air than when the sun is overhead.[2] In the extreme situation of dense fog, the intense scattering by water vapour molecules blurs a source of light and turns it into a diffuse source. This situation is well known to mountain climbers from whiteouts on glaciers or snowfields, where the whiteness of snow and ice combined with the extremely diffuse light and the complete absence of shadows makes the scene being perceived as flat. The same scattering by water droplets can be observed when an intense beam of light illuminates fog (Fig. 7.2).

At higher and higher elevations in the atmosphere, there are less and less water vapour and particulate matter and the sky looks less hazy and bluer. In the absence of scatterers, Mie scattering does not take place and Rayleigh scattering, which is very effective for blue light, dominates. This is the reason why the sky looks bluer and bluer at higher elevations, and it is darker looking straight up than looking

---

[2] One consequence is that the light rays are refracted (bent) more than the rays coming from overhead. When we see the sun setting on the horizon, it is already below it.

Fig. 7.2   Scattering by water droplets in the air in morning fog.

toward the horizon. For the same reason, in polar regions where there is essentially no water vapour and there are no large particles in the air, the sky looks clean and blue, light sources are sharp and not blurred, and one can see very far.

At higher and higher elevation there is less and less air and the rarefaction means that there are less molecules available for the Rayleigh scattering. Then the sky looks darker and darker, as on a clear day on very high mountains. Going higher to outer space, the sky appears completely black.

7.3   Put your sunscreen on

The sun emits electromagnetic radiation with a blackbody spectrum, which means that more energy is emitted around a central wavelength (500 nm, corresponding to yellow-green colour) than around any other wavelength. Very long (near the infrared region) and very short (in the ultraviolet region) wavelengths are emitted comparatively very poorly, meaning that a very small fraction of the total energy is emitted in these regions. Here we are concerned with emission in the ultraviolet (UV) region of the electromagnetic spectrum, where the radiation output from the sun is small but non-zero. The ozone layer present in the atmosphere at elevations between 20 km and 30 km absorbs most of this radiation, but some still arrives to the surface of the Earth. Mountaineers go high above sea level, where UV rays have travelled less atmosphere than when they arrive to sea level, and so these climbers receive higher UV doses than if they stayed lower down, especially if they stayed inside their homes. Another reason for increased radiation is the reflection from snow and ice. Apart from sunburns, a tangible effect of ultraviolet rays is evident on the flys of tents which are pitched at high elevation for long times.[3]

The effect of UV rays on biomolecules is very negative. Photo-damage mutates genetic material and kills cells, the chromophores of DNA being prime targets. When it comes to multi-cellular organisms, UV rays can damage important parts. The effect of large doses of UV rays in humans is skin cancer. Melanoma can be, or become, invasive and travel inside the body far from the skin. A layer of material absorbing the UV rays can avoid this, or largely reduce the danger. The first protection for human life is, of course, the ozone layer in the atmosphere, which absorbs radiation with short wavelengths ($\lambda < 295$ nm) because these UV photons excite an optical transition in ozone. The ozone layer stretches vertically for kilometers in the rarefied conditions present above 20 km of elevation. If this layer were to be brought to "standard" conditions of temperature and pressure (*i.e.*, zero degrees Celsius and one atmosphere), it would have a thickness of only 0.3 cm, yet it removes most of the UV radiation that would otherwise reach the Earth. This is why the destruction of the ozone layer by chlorofluorocarbons (CFCs) was such a serious problem and why stopping the emissions destroying this thin and fragile layer, and monitoring its conditions, are extremely important tasks in environmental science [Boeker and van Grondelle (2011)].

---

[3]For the same reason, ropes should be sheltered from UV rays when not in use.

The UV rays generated from the sun, however, are not stopped completely by the ozone layer. The second layer of protection can be, of course clothing, and the third layer of protection, which becomes the second one for exposed skin on the face or hands, can be sunscreen with a very high protection factor.

To appreciate the effect of a protective layer, and its consequences if it is absent, it is useful to understand the basics of the propagation of electromagnetic radiation through a medium. As a beam of radiation (for example UV rays) travels through a material (for example a layer of sunscreen), its intensity I (energy crossing the unit of area perpendicular to the flow per unit time, or flux density) decreases. The law describing this attenuation with the distance travelled is used in many areas of physics, engineering, and biology, in which it receives various names. Say that the beam travels a slab of material along a straight axis and that z is the distance travelled along this axis, while $I(z)$ is the intensity of the beam at the position z. The attenuation is exponential, as described by the empirical law

$$I(z) = I_0 \, e^{-kz},$$

where $I_0 = I(z = 0)$ is the intensity of the beam upon entering the slab at $z = 0$ and k (which has the dimensions of an inverse length, $[k] = [L^{-1}]$) is called *absorption coefficient*. In environmental physics the law is called *Lambert–Beer–Bouguer law*. It is important to realize that this is a macroscopic, phenomenological description. The loss of energy by the radiation as it travels is due to its interaction with the atoms and molecules of the medium in which it propagates. This interaction is a quantum phenomenon [Griffiths (2005); Gasiorowicz (2003); Schiff (1968); Messiah (1961)]: photons excite electrons in atoms or molecules and make them jump from one energy level E_1 to a higher energy level E_2, but only if the frequency of the radiation is right, that is, close to the resonant frequency

$$\nu_{12} = \frac{E_2 - E_1}{h}$$

determined by energy difference between these two energy levels and by the Planck constant h. Some frequencies are "just right" and photons are absorbed by the quantum system (atom or molecule), while photons of other frequencies are not absorbed at all [Griffiths (2005); Gasiorowicz (2003); Schiff (1968); Messiah (1961)]. At the macroscopic level, this fact is reflected in the property of the absorption coefficient k, and of the attenuation of the beam, of being strongly dependent on the frequency ν of the radiation.

Continuing, as the beam travels through a medium composed of atoms or molecules which do absorb photons, the longer the distance z travelled, the more absorption events occur, and the more energy is taken away from the beam. As a result, the intensity I of the beam decreases, and the particular features of the process determine the exponential character of the Lambert–Beer–Bouguer law. As already said, this is an experimental law, which is not difficult to verify in the lab. However, it can also be derived theoretically. The derivation (*e.g.*, Boeker and van Grondelle (2011)) requires an understanding of the quantum processes of absorption, which is now standard textbook material due mainly to work by Albert Einstein [Griffiths (2005); Gasiorowicz (2003); Schiff (1968); Messiah (1961)]. In addition, when a beam travels a macroscopic distance which, depending on the context, could be microns, millimeters, or kilometers, it encounters many atoms. There is an enormous number of atoms or molecules in any macroscopic sample. For example, it is well known from high school chemistry courses that a mole of a substance (a quantity equal to its atomic or molecular weight expressed in grams) contains an Avogadro's number of these objects,[4] that is $N_A = 6.022 \cdot 10^{23} \, \text{mol}^{-1}$. Because one cannot possibly calculate and describe all these particles' dynamics and their interactions, one replaces the real particles with an average fictitious atom or molecule and uses the methods of the area of physics known as statistical mechanics [Carter (2001); Fermi (1956); Sears and Salinger (1975); Zemansky and Dittman (1997)]. By combining quantum mechanics

---

[4]Incidentally, it was Einstein again who gave a major contribution to a relatively precise determination of Avogadro's number through his famous 1905 study of Brownian motion [Einstein (1905)].

and statistical mechanics it is possible to derive theoretically the Lambert–Beer–Bouguer law.

There is a slightly different mathematical way to express the macroscopic Lambert–Beer–Bouguer law. Instead of using the base e for the exponential, the base 10 is often preferred by experimental scientists, who rewrite this law as

$$I(z) = I_0 \cdot 10^{-\tau} = I_0 \cdot 10^{-\epsilon C z},$$

where I and z have the same meaning as before and now the dimensionless *optical distance*

$$\tau = \epsilon C z$$

is composed of:

- the *molar extinction coefficient* ϵ, measured in units $\frac{\mathrm{dm}^3}{\mathrm{mol} \cdot \mathrm{cm}}$, which summarizes the microscopic processes of energy absorption;
- the concentration of absorbers C, which is measured in $\mathrm{mol}/\mathrm{dm}^3$.

This form of the law is more useful for quantifying, for example, where (that is, at which length z travelled) the intensity I of the beam is reduced to one tenth of its original value I_0. In order to relate the quantities k to ϵ and C, one equates the two expressions of the Lambert–Beer–Bouguer law, which give the same intensity $I(z)$. The initial intensity I_0 cancels out and one is left[5] with $e^{-kz} = 10^{-\epsilon C z} = e^{-\epsilon C z \ln 10}$. Then the arguments of the two exponentials are equal and one obtains

$$k = \epsilon C \ln 10.$$

A plot of I/I_0 as a function of τ is shown in Fig. 7.3. As can be seen from this plot, an essential feature of the Lambert–Beer–Bouguer law is that the attenuation of the beam is exponential, that is, very fast.

The absorption coefficient k has the meaning of inverse of the length scale over which the attenuation occurs. In fact, if the beam travels a distance z_* equal to the inverse of the absorption coefficient,

---

[5]We use the well known property $a^x = e^{x \ln a}$ for any $a > 0$.

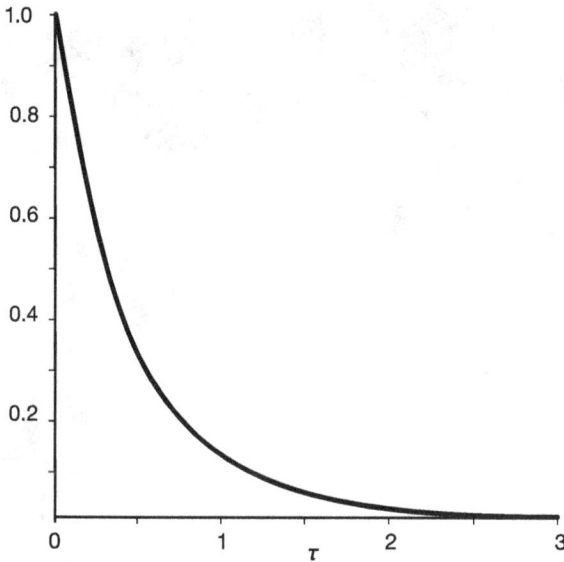

Fig. 7.3 Exponential attenuation in the Lambert–Beer–Bouguer law.

it is $kz_* = 1$ and the beam intensity is attenuated by the factor

$$\frac{I\left(z = k^{-1}\right)}{I_0} = \mathrm{e}^{-1} \simeq 0.3679.$$

This means that, when the beam has travelled a few length scales k^{-1}, the attenuation is almost complete. It is easy to understand, therefore, why even a relatively thin layer of material designed to absorb radiation in a certain range of frequencies, such as sunscreen, can be very effective. Put your sunscreen on.

7.4   Lichen it or not

Let's face it, lichens grow on alpine rocks, including those on our rock routes, lichen it or not. Most often we observe circular patterns of lichens which evolve from a point, which becomes a centre, and then expand radially along the surface of the rock (Fig. 7.4). Circular patterns starting from different centres grow and sometimes they meet and intersect.

Fig. 7.4   Lichens growing on alpine rock.

Lichens growing on alpine rocks can have beautiful colors and, in some cases, cover much of the rock surface and even entire walls. The description of their growth can be quite complicated and is similar, in many aspects, to the growth of fairy rings of mushrooms found in the meadows below the alpine. Fairy rings were once attributed to magics, and stepping inside a fairy ring of mushrooms was believed to bring bad luck. They are mentioned in Shakespeare's *The Tempest* and *A Midsummer Night's Dream* and they sprout in much literature and folklore, especially those from the British Isles, which means that people were wondering about them long before science was invented. Often one sees rings of dead grass with lush grass on the outside. After a good rainfall, mushrooms sprout on the edge, acting as markers for the mycelium that grows underground. The kinetics of the expansion are a fascinating topic in non-linear dynamics. The mycelium grows radially outwards as it feeds on organic matter contained in the soil. There is a front that expands outwards and leaves behind it dead mycelium, which decays and acts as a fertilizer for grass. Strangely, also the advancing front contains extra nutrients pre-digested with enzymes which cause extra grass growth in front of the ring. The expansion rate v of the ring is measured to be approximately 0.1–0.3 meters/year. Since some fairy rings with diameters d of hundreds of meters are observed, they must have an age, say, $t = \frac{d}{2v} \simeq 200\,\text{m}/0.2\,\text{m/yr} \sim 1000\,\text{yr}$. While this estimate may be too generous, 500–600 years is a realistic estimate for the age of these large fairy rings.

When different rings of the same species encounter, they form cusps in the region of intersection. In these intersection regions, the mycelia of different rings compete for nutrients. When each mycelium reaches the dead zone of the other one, it stops growing.

Growth, morphogenesis, and pattern formation are important topics in non-linear science, with many applications to biology. Reaction-diffusion was suggested as a tool to model morphogenesis by the famous mathematician, computer scientist, theoretical biologist, and philosopher Alan Turing [Turing (1952)].

Mathematical models of the growth of fairy rings were developed in the 1950s and 1970s [Parker-Rhodes (1955); Stevenson and Thompson (1976); Scott (2005, 2007)]. The growth happens in two spatial dimensions and the growth rate is, typically, a function of time. The mathematical model, therefore, consists of a system of partial differential equations in two dimensions which describe simultaneously the fungi and their nutrient. Such a system is of the non-linear reaction-diffusion type, which means that a diffusion phenomenon is accompanied by a source for the quantity that is diffusing. Diffusion is very common in nature, for example the propagation of heat by conduction obeys the diffusion law, as well as the diffusion of particles in a still medium in the absence of currents, for example the slow diffusion of a drop of colored ink in a glass of still water. The diffusion equation is a prototype of partial differential equations.[6]

---

[6]Most phenomena studied in physics are described by partial differential equations of second order. From the physical point of view, most second order partial differential equations encountered in textbooks and in mathematical modelling can be classified in three types which describe, respectively, wave phenomena (hyperbolic type), diffusion processes (parabolic type), or static phenomena (elliptic type).

In two dimensions, using as coordinates in the plane the distance r from a fixed origin to move to circles of larger and larger radii and an angle θ to move along circles of fixed radius, the diffusion equation for a quantity $u(t, r, \theta)$ which depends on time and position, takes the form

$$\frac{\partial u}{\partial t} = D \left(\frac{\partial^2 u}{\partial r^2} + \frac{1}{r} \frac{\partial u}{\partial r} \right).$$

Here we have assumed cylindrical symmetry, *i.e.*, all the directions in the plane are equivalent and the dependence on the coordinate θ drops out. The quantity in brackets on the right hand side of this equation, denoted by mathematicians with $\nabla^2 u$ (the *Laplacian* of u, see Appendix A), is essentially the spatial curvature of the function u, so the equation says that the time rate of diffusion is proportional to the spatial curvature of the function describing the concentration of the quantity that is diffusing. The more pronounced this curvature, the faster the diffusion process. For a very flat configuration, with very small gradients $\vec{\nabla} u$ and little curvature $\nabla^2 u$ in the concentration u of the diffusing quantity, diffusion is very slow and it stops altogether if the concentration becomes uniform, corresponding to $u = $ constant and $\vec{\nabla} u = 0$. In this case equilibrium has been reached and the situation becomes rather boring. *Vice-versa*, when there are large gradients $\vec{\nabla} u$ and large curvatures $\nabla^2 u$, diffusion is rapid.[7]

Here we assumed that the quantity D, the *diffusion coefficient*, is constant. Everything else equal, the larger the value of D, the faster the speed of diffusion $\partial u / \partial t$.

Reaction-diffusion is a more complicated phenomenon: in addition to diffusion, there is a source of the quantity u (for fairy rings of mushrooms, the nutrients), which is not prescribed *a priori*, but depends on the quantity u itself, much like two reactants in a chemical reaction. This more complicated reaction-diffusion phenomenon

---

[7]Relatively speaking, the diffusion coefficient D is usually small for most natural processes, making diffusion a slow process compared, for example, with transport by a current or a wind. It is still an important process when small distances are involved, for example in transport across biological membranes.

in two spatial dimensions is described by an equation of the type

$$\frac{\partial u}{\partial t} = D \left(\frac{\partial^2 u}{\partial r^2} + \frac{1}{r} \frac{\partial u}{\partial r} \right) + f(u),$$

where $f(u)$ is an appropriate non-linear function of the unknown u itself. This type of equation is common in many natural and man-made phenomena, including chemical reactions, propagation of electric pulses down nerves, collective phenomena such as the synchronization of the light flashes of Asian fireflies, flight formation of birds, and the schooling of fish (see, *e.g.*, [Scott (2007, 2005)] and the references therein). Because of their non-linearity, reaction-diffusion equations cannot be solved analytically and require numerical solution with computers.

The observed features of fairy rings of mushrooms are modelled well by reaction-diffusion mathematical models. The mathematical modelling is useful for managers of golf courses who want to limit, or eliminate, fairy rings of mushrooms and make their grass perfect. However, there are more useful applications of these mathematical models. Other systems exhibit the same kind of growth, for example certain species of saltbrush that grow in Australia. This saltbrush is an important source of food for sheep in the semi-arid landscape of Australia, and there their modelling becomes more important for the economy [Scott (2005)].

Now, back to our lichens. Lichens result from the composition of an alga or cyanobacterium with a fungus, living in symbiosis. There is a difference with respect to fairy rings: lichens produce their own food by photosynthesis contrary to fungi which do not photosynthesize but absorb nutrients from their environment. As a result, lichens do not leave behind a dead area when they expand circularly.

Lichens typically grow slowly with a rate $v < 1$ mm/yr, but some species grow at the exceptional rate of 0.5 m/yr. Most, but not all, lichens found on alpine rocks are of the crustose type, made of flakes that grow on the rock surface and resemble peeling paint, hence their name. Since the growth of some lichens is very slow and regular, they can be used to date events, giving rise to *lichenometry*. Lichenometry is a method of geochronological dating that assumes constant rate of radial increase with a specific value and uses the growth of lichens to determine the age of an exposed rock. The method can work for the time that lichens grow on exposed rock, up to ten thousand years.

The growth of lichens is also described by the reaction-diffusion paradigm in several mathematical models [Topham (1977); Proctor (1977); Aplin and Hill (1979); Childress and Keller (1980); Hill (1981)]. The biology of the growing process is not completely known yet and the mathematical models also have great richness to explore.

7.5   Places with a lot of potential

Being in the mountains means being high up and close to the sky and the clouds. For some, this may be a zen-like situation with a high potential for enlightenment but also for something else. Being high up is not a good thing when the weather changes and the clouds become electrically charged. Lightning strikes hit the mountains very often and then these become extremely dangerous places. The media report that, on average, the probability of being hit by lightning is very low, something like 1 in 400,000 for each year of life. This means nothing for a mountaineer because this probability is estimated for average citizens who spends most of their life at lower elevations, inside safe buildings, and likely in a city with lots of shelter, given that in the Western world most of the population is concentrated in cities. The probability of being hit by lightning is much, much closer to 1 (*i.e.*, certainty) when one is an exposed place close to the clouds and without shelter. Indeed, lightning is one of the first few causes of death among mountain climbers. Once I was caught out and almost zapped near a summit. Since then I have been very scared of thunderstorms and I have become very careful in the mountains. This caution didn't avoid repeating the experience twice years later.

In many mountain areas of the world, the afternoon heat thunderstorm is a classic summer feature. Hot air rises, clouds become electrically charged, and thunderclouds build up. Generally, in Colorado, in Wyoming, or in the Italian Dolomites one should be off the peaks by 1 pm. If we see that the thunderstorm is on its way to us, we should descend as quickly as possible. If we see lightning and hear thunder on a climb, how do we assess the danger? We can get a rough estimate of the distance from a lightning bolt by counting the time Δt in seconds between seeing the flash and hearing the thunder. The thunder is the sonic boom caused by the rapidly expanding plasma ionized by the giant spark. The temperature in the lightning channel can reach a peak value of 20,0000–30,000 K, as measured by spectroscopy. The speed of sound at sea level, zero degrees Celsius, and pressure of one atmosphere, is about 330 m/s, which we will take to be indicative also of the sound speed in the mountains. The distance travelled by sound between the flash and the thunder is $d = v\Delta t$. Say that we counted ten seconds between lightning flash and thunder, then the bolt hit approximately $d \simeq (330\,\text{m/s})\,(10\,\text{s}) = 3.3\,\text{km}$ away. The rule of thumb is a kilometer every three second delay. If the time Δt becomes something like one second or less, we are in serious trouble.

The best thing to do with thunderstorms, of course, is not to get caught, to keep watching the weather and its changes, and to descend immediately or seek a proper shelter well ahead of the storm. This is easier said than done, for sometimes one does not see the weather approaching from the opposite side of a mountain or one is in a place from which retreat is difficult and long, such as a large rock wall, but a little thought can go a long way and some knowledge of electrical phenomena may be of some help. The physics of a lightning flash is not simple [Rakov and Uman (2007)], but one can try to reduce the danger. Things that we can do include obvious ones such as not camping in a place exposed to lightning; if camping, avoiding to lie down during a thunderstorm to prevent ground currents from going through the body — sitting with feet together, hands around the head and elbows on our knees. Other things we can do include being aware of signs of static buildup on a climb, some of which are obvious and some come too late, such as crackling sounds, sparks on ice axes, a strange buzz in the air, hair standing straight, tingling sensations.

The cause of an electric discharge is the difference in the electric potential V at two points [Griffiths (1999); Reitz and Milford (1960); Jackson (1975)]. This difference is caused by the bottom of a cloud becoming electrically charged with negative charge. Unbalanced charges in one area create a difference of electric potential between this area and neighbouring points [Griffiths (1999); Reitz and Milford (1960); Jackson (1975)]. The electric potential $V(\vec{x})$ is a scalar function of the position $\vec{x}$ and, in dynamical situations, it also changes with time. When the electric potential, which is electric potential energy per unit charge, is very different at two nearby points, meaning that large accumulations of charges have occurred, a large electric field is present between them. In fact, the electric field, a vector, is (minus) the gradient of the electric potential [Griffiths (1999); Reitz and Milford (1960); Jackson (1975)],

$$\vec{E} = -\vec{\nabla}V.$$

The minus sign is due to convention and is not crucial: it says that the electric field points from positive to negative charges, or from positive charges to spatial infinity. The value of the potential V at a point is also not important, since the potential is defined up to an arbitrary constant. What matters are *differences* in the potential. The electric force acting on an electric charge q is $\vec{F} = q\vec{E} = -q\vec{\nabla}V$, so the larger the gradient of the electric potential V, the larger this electric force. Potential differences of thousand of Volts rip electrons from (normally neutral) atoms and ionize them, generating sparks a few centimeters long. Free electrons and ions in the air conduct electricity. Lightning bolts in the atmosphere involve potential differences of many million Volts and large electric fields creating huge currents of thousands or hundreds of thousands of Amperes, which have a devastating impact on the media in which they travel. A leader of negative charge moves from a negatively charged cloud to the ground and usually positive charge moves upward in a return stroke. Lightning bolts can occur inside a cloud, between clouds, between a cloud and the ground, and in any gap between rocks or the ground which are travelled by the current. A lightning bolt travels at very high speeds, with the leader being much slower than the return stroke, and it can split or generate side bolts.

If a group is caught in a thunderstorm, its members should separate to reduce the number of possible casualties and so that survivors can assist potential victims. Entire groups of people or animals have been hit by lightning strikes, so one should not be afraid of separating from a group, while keeping in sight. A hiker was killed on an insignificant grassy mountain top that I used to visit frequently and, on another occasion, some twenty sheep were killed together in the same place. Contrary to popular saying, lightning does strike repeatedly in the same places and knowing these places is very useful.[8] On August 26, 2016, 323 reindeer were killed together in a thunderstorm on the Hardangervidda Plateau of Central Norway. This large number is explained by the fact that reindeer tend to group very closely, probably more so if they are scared by a thunderstorm.

Lightning bolts are notoriously bizarre, so one should be aware of direct strikes, side strikes, and ground currents.

To avoid direct strikes, we should avoid obvious danger areas, summits, ridges, rock faces, dihedrals, chimneys, open meadows, lakes, and lone trees or spires. One should get away from exposed places fast. On ridges, one should go down one side. Bolts travel along faces and caves should be avoided. Overhangs are not safe, either. The usual recommendation is to stay within the forty-five degree shadow of a tall object, preferably the tallest around, which will act as lightning rod. Being closer to it may result in a side spark from the object, the equivalent of touching a lightning rod, or in very strong ground currents. If lightning bolts are really close, one should discard temporarily any heavy metal gear: ice axe, crampons, rock hammer, piles of rock gear and carabiners, deadmen, snow stakes, and cooking gear have to go. Also long and pointy items such as skis, ski poles, hiking poles, avalanche probes, and glacier wands must be dropped since points are places of very high curvature and electric charge density, and they easily emit sparks which ionize the air and, in critical conditions, may create an easier channel for an electric discharge to travel in.

If a person is hit directly by a lightning bolt it probably means death, but there are also side strikes, gap sparks, and ground currents that have been survived and one can try to minimize their damage.

---

[8]In Europe, where records have built up for many centuries, certain toponyms give away the dangers of particular locations.

It is useful, in particular, to know something about ground currents when looking for safety and choosing a "best place" in which to stop during a thunderstorm. Ground currents, resulting from large potential differences in the ground, but not as large as those causing the main strike, can be very strong: they sometimes carve trenches in the soil where they travel. A potential difference of many thousands of Volts can be created across the space of a few tens of centimeters, that is, between one's feet or between different parts of one's body if lying down. Because of the possibility of these intense ground currents, if caught in a thunderstorm, we don't spread our body on the ground and we don't keep our feet apart. It is best to squat low but not to sit down, with feet together, hands over the head, and elbows touching the knees. This position minimizes currents through vital organs by creating paths for the currents that run outside of them.

It is a good idea to insulate from the ground as much as possible, and to allow outer clothing to get wet, which facilitates an effect in which currents travel more on the outside of a conductor, in this case the human body, than inside it, which would have catastrophic effects on the heart, brain, and internal organs (this observation was made already by Benjamin Franklin [Krider (2006)]). There have been situations in which people hit by lightning have suffered severe burns on the outside of their body but no internal injury, which would have been fatal. Obviously one should avoid touching, or being near good conductors or metal. The steel cable on a via ferrata is exactly what one should avoid in a thunderstorm, and many fatalities related to lightning in the European Alps occur on via ferratas. A friend of mine was zapped by lightning on a via ferrata. He was rescued promptly but spent a terrible time being completely paralyzed and not knowing if this situation was going to be permanent. After an hour or so, he started to function again and, surprisingly, he did not suffer any damage. He was lucky but other people whose bodies were traversed by primary or secondary currents have not been as lucky.

There are many bizarre and terrifying stories about lightning. A very strange one is that Benjamin Franklin, who used to fly kites during thunderstorms in order to study electricity, somehow survived without incidents [Krider (2006)]. His colleague Georg Wilhelm

Richmann from St. Petersburg attempted to perform experiments during a thunderstorm in 1741 and was killed by ball lightning. Ball lightning, known to sailors as St. Elmo's fire, is a very difficult phenomenon to model and to explain physically. A glowing ball in the air rolls and moves around and sometimes explodes and disappears. There are several stories in which people were not harmed by ball lightning touching them, but Professor Richmann was killed instead.

Lightning bolts can follow bizarre and unpredictable paths. There are stories of lightning traveling inside houses, choosing strange paths, destroying parts of them and leaving small animals unscathed nearby, then exiting and killing larger animals. There are even stranger reports of images being burned on the inside of the skins of animals killed by lightning. In 1812 six sheep were struck dead near Bath, England and, reportedly, on the inside surface of the skin of each one of them there was an image of parts of the surroundings [Monthly Meteorological Magazine (1883)]. The skins were exhibited in Bath for some time. There are many accounts, some by Benjamin Franklin himself, of "photographic images" caused by lightning. Most of these stories are about thunderstorms and lightning striking at lower elevation or at a sea level, and they are horrific as they can be bizarre. Even more scary are the stories of mountaineers caught by electric storms in exposed places at high elevation. High places have a lot of potential, including electric potential (differences).

7.6   Science labs in the mountains

To finish our excursion on physics in the mountains, let us mention science facilities actually located on alpine peaks. Many early mountaineers were scientists and they took on their climbs instruments to measure atmospheric properties. Later, it was natural to build meteorological stations at higher elevation on mountains. But, even though meteorological phenomena govern the life of people living in the mountains and can be matters of life and death for mountaineers, the days in which local atmospheric studies were forefront science are long gone. However, there are still many science facilities devoted to the pursuit of fundamental science which need to be built on, or inside, mountains.

First of all astronomical observatories, which investigate celestial objects and the birth and evolution of the universe itself and of the structures in it, are built on mountain tops for various reasons. On a high mountain the light reaching the telescope traverses a shorter distance in the atmosphere than if the telescopes were located lower down or at sea level, and the winds in the atmosphere cause scintillation of the images. Even little kids know that stars twinkle. This fact is evident by looking at a star from low elevations: it twinkles due to the motions of the air interposed between the telescope and the light rays arriving on the Earth, and this twinkling does not bode well for obtaining the sharp images needed by astronomers. Reducing the length travelled by the light in the atmosphere reduces the problem, and the densest layers of air are the bottom ones (Sec. 5.2).

Second, at higher elevation there is less water vapour and the atmosphere is more rarefied, which means less scattering of light by molecules of the atmosphere and less blurring of the images. It would be a waste to spend many millions of dollars to build a telescope which can achieve a high resolution and then install it in a place in which there is large scintillation or the sky is not dark: here is where mountains are very useful to this science.

Third, mountain tops are further away than lower elevations from man-made lights which ruin the observations of very dim celestial objects and compromise the darkness of the sky (*light pollution*).

Further, as seen in Sec. 7.2, the water vapour, aerosols, and particulate matter in the atmosphere scatter light, and they are mostly concentrated at lower elevations. Building observatories on remote high mountains with dry clean air reduces this factor.

To begin with, many older European mountaineers have stumbled on the Gornergrat infrared observatory, which was located at 3,120 m on the Gornergrat ridge near Zermatt, Switzerland. It hosted a 1.5 m infrared telescope until it was decommissioned in 2005. On this site there are now a hotel and a planetarium for the popularization of astronomy. Stepping up in size, several telescopes are located on the Teide volcano on the Spanish island of Tenerife, off the coast of Africa, where light pollution is minimal. The largest telescopes on Earth are located on higher mountains: many are housed on Mauna Kea in Hawaii, and many more have been built on the arid mountains of Northern Chile since the 1960s. The ideal observing conditions and possibly the best skies on Earth found in Northern Chile have motivated the construction of the Atacama Pathfinder EXperiment [APEX (2017)], the Paranal [Paranal (2017)], Cerro Armazones [Cerro Armazones (2017)], Las Campanas [Las Campanas (2017)], La Silla [La Silla (2017)], Cerro Tololo [Cerro Tololo (2017)], and Atacama Large Millimeter Array [ALMA (2017)] observatories there. These mountains host projects such as the Very Large Telescope [VLT (2017)], the Gemini telescope [Gemini (2017)], the Large Synoptic Survey Telescope [LSST (2017)], and the future 42 m European Extremely Large Telescope [ELT (2017)]. The dry air and remoteness make these mountain sites irreplaceable for astronomy. It is interesting that astronomical tourism has developed in the wake of big astronomy and that there are now several smaller observatories devoted specifically to astronomical tourists.

Moving on, there are physics experiments actually performed on, or inside, mountains and requiring the mountain as an essential part of the facility.

A famous experiment in the history of physics confirmed a prediction of Einstein's theory of Special Relativity [Landau and Lifshitz (1989); Rindler (1991); Fayngold (2008); Takeuchi (2010); Dragon (2012); Faraoni (2013); Rafelski (2017)] and was performed on Mt. Washington, New England by the physicists Bruno Rossi and David Hall in 1941 [Rossi and Hall (1941)]. The experiment verified the relativity of time for different observers and the time dilation formula predicted by Special Relativity [Landau and Lifshitz (1989); Rindler (1991); Fayngold (2008); Takeuchi (2010); Dragon (2012); Faraoni (2013); Rafelski (2017)] by measuring the lifetime of particles called muons present in cosmic rays. A muon ($\mu^{\pm}$) is an unstable particle which can have positive or negative charge and decays into electrons (e^-) or positrons (e^+, the antiparticle of the electron), neutrinos (ν) and their antiparticles, the antineutrinos ($\bar{\nu}$) [Griffiths (2008); Cottingham and Greenwood (2003)]. The muon decays are described by

$$\mu^- \longrightarrow e^- + \nu_\mu + \bar{\nu}_e,$$
$$\mu^+ \longrightarrow e^+ + \bar{\nu}_\mu + \nu_e,$$

where $\nu_{e,\mu}$ denote the electron and muon neutrinos, respectively.[9] In a reference frame in which it is at rest, a muon decays in $\tau = 2.2$ μs, the *proper lifetime* of the muon. Cosmic rays from outer space hit the nuclei of atoms in the upper troposphere and produce muons as secondary cosmic rays. These particles travel at speeds v near the speed of light. It is observed that the produced muons reach the surface of the Earth before they decay, which means that in the reference frame connected with the ground in which they move at $v \approx c$, their lifetime is much larger than 2.2 μs. In fact, if their lifetime was 2.2 μs, they would travel only $(2.2 \cdot 10^{-6}\text{s}) \cdot (3 \cdot 10^8 \text{ m/s}) \approx 660$ m. Instead, the upper atmosphere where these particles are created by primary cosmic rays reaches an elevation of ~ 10 km and the muons are detected much lower down on the surface of the Earth. One concludes that the lifetime of the muon measured in a frame of reference in which it is in motion is longer than the lifetime measured in a frame of reference in which it is at rest. This *time dilation* is predicted by Special Relativity [Landau and Lifshitz (1989); Rindler (1991); Fayngold (2008); Takeuchi (2010); Dragon (2012); Faraoni (2013); Rafelski (2017)]. If $\Delta\tau$ is a time interval measured by an observer (we say that his clock is at rest) and Δt is the corresponding time interval measured by an observer moving with speed v with respect to the former, the relation between the two measurements is [Landau and Lifshitz (1989); Rindler (1991); Fayngold (2008); Takeuchi (2010); Dragon (2012); Faraoni (2013); Rafelski (2017)]

$$\Delta t = \frac{\Delta\tau}{\sqrt{1 - v^2/c^2}}, \tag{7.1}$$

---

[9]Light elementary particles called *leptons* come in only three families or *flavours*, the electron, muon, and tau families [Griffiths (2008); Cottingham and Greenwood (2003)].

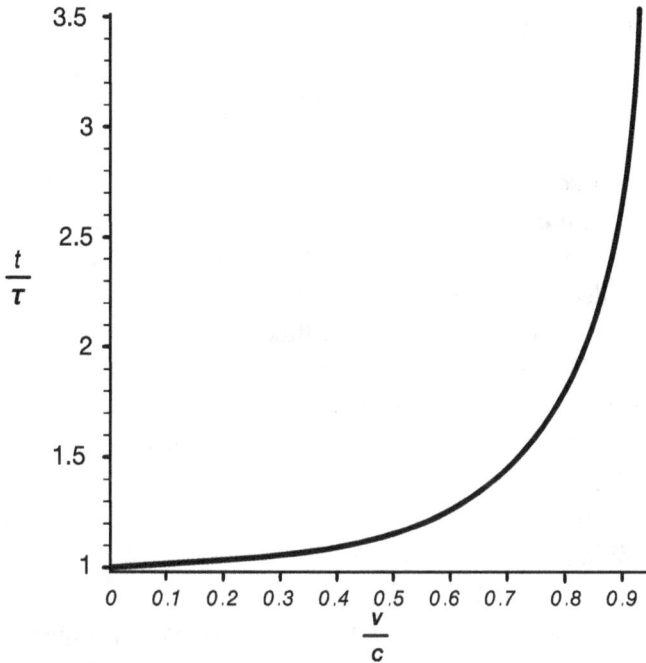

Fig. 7.5   The ratio between coordinate and proper time intervals for the muon.

where c is the speed of light *in vacuo*. The curve describing the ratio $\Delta t/\Delta \tau$ between the time interval in the frame in which the muon is moving to the one in which the muon is at rest is plotted in Fig. 7.5 as a function of the muon speed, measured in units of the speed of light c. This curve becomes very steep as v approaches c and it diverges[10] as $v \to c$.

This time dilation effect stretches the muon lifetime for an observer in an inertial frame attached to the Earth, in which the muon goes by with speed v. In this frame, the typical time taken by

---

[10]The muon speed can become arbitrarily close to c but it can never reach it because this particle has a non-zero mass [Landau and Lifshitz (1989); Rindler (1991); Fayngold (2008); Takeuchi (2010); Dragon (2012); Faraoni (2013); Rafelski (2017)]. Only particles with zero rest mass can travel exactly at the speed of light.

a muon to decay (muon lifetime) is

$$t_{\text{decay}} = \frac{\tau}{\sqrt{1 - v^2/c^2}}.$$

Let us insert some numbers in this formula to get a feeling of what's involved. If the muon travels at 99% of the speed of light, $v = 0.99\,c$, then $t_{\text{decay}} = 16\,\mu$s in the Earth's frame. Correspondingly, the distance travelled by this muon is $\left(16 \cdot 10^{-6}\ \text{s}\right) \cdot \left(3 \cdot 10^8\ \text{m/s}\right) \approx 5000$ m.

Rossi and Hall measured muon fluxes at the top and at the bottom of Mt. Washington in New England, 2000 vertical meters apart. A more precise experiment performed at CERN in 1971 with muon speed $v = 0.9994\,c$ [Bailey *et al.* (1977)] verified the time dilation formula (7.1) with better precision. The version of the experiment requiring a mountain trip is no longer frontier research but is nowadays possible as a training experience for undergraduate students in physics [Easwar and MacIntire (1991)].

There are more modern fundamental physics experiments requiring a mountain. In the Apennine mountains of central Italy, under Gran Sasso, a tunnel shelters experiments from cosmic rays. If we are trying to detect an elusive particle which travels through space and interacts very little and very rarely with the atoms of our particle detectors, the surface of the Earth is not a good place to be. Given long observation times, very sensitive particle detectors could perhaps reveal the rare interaction between these particles and the nuclei in the atoms of ordinary matter, but these rare events are lost in the noise of much more frequent signals caused by cosmic rays interacting with the same atoms of the detector. In order to even begin such an experiment, one needs to shelter it very carefully from cosmic rays causing all these spurious detection signals. What better place than inside a mountain? Under Gran Sasso, a tunnel built into the mountain provides exactly such a place. We now know that most of the matter in the universe is not of the kind that we are made of. In fact, it has been known for decades by studying the dynamics of galaxies that most of the mass in galaxies is not luminous, as it would be if it was made of baryons, *i.e.*, protons and neutrons, which compose ordinary nuclei and atoms. Instead, most of the galactic mass must be made of some exotic non-luminous form of matter, which does not interact through the electromagnetic force, one of only four fundamental interactions.[11] This missing mass, however, gravitates and has been called *dark matter* [Binney and Tremaine (1987)]. Evidence for its existence comes from the study of independent astrophysical phenomena such as the trajectories and speeds of stars in galaxies, and the bending of light by galaxies which act as lenses for light rays coming from further away by bending them (*gravitational lensing*) [Schneider, Ehlers, and Falco (1992)]. There is evidence for dark matter also at the much larger spatial scale of galaxy clusters: the intergalactic medium is much too hot (it emits in the X-ray band!) to be moving in the gravitational field of visible matter alone. To achieve the observed speeds and temperatures, it is necessary that there is much more (dark) matter in clusters than the observations of visible matter can account for. This dark matter does not emit light, but it gravitates and it accelerates the ordinary matter particles of the intergalactic medium to high speeds, causing them to emit X-rays. There is plenty of further evidence for dark matter, or something that causes the same effects, from the bending of light in galaxy clusters, too. Further, in cosmology, in order to fit models to the very precise data from temperature fluctuations of primordial origin in the cosmic microwave background [Durrer (2008)], it is required that a considerable part of the energy content of the universe be in the form of dark matter, or something which behaves in the same way [Particle Data Group (2017)].

Dark matter is now a paradigm of cosmology and of the astrophysics of galaxies and galaxy clusters (see [Bertone, Hooper, and Silk (2005)] for a review). Particle physicists have come up with candidates for dark matter in the form of exotic particles which interact extremely weakly with ordinary matter, as required in order to explain the details of the galactic dynamics. These particles are

---

[11]We know only four fundamental interactions: gravity (well known to mountaineers), electromagnetism (which describes light and optical phenomena and, through the attraction and repulsion of electric charges, is responsible for the existence of atoms and molecules), the strong nuclear force (which keeps nuclei together), and the weak nuclear force (responsible for radioactivity).

predicted in a theory (*supersymmetry*) which is speculative and goes beyond the known and well-tested Standard Model of particle physics [Weiss and Zumino (1974a,b); Aitchison (2007); Baer and Tata (2006); Binétruy (2006); Labelle (2010)]. Thus far, however, nobody has ever observed directly such a dark matter particle in a lab. There are now several experiments aiming at detecting dark matter particles directly in the laboratory, and the Gran Sasso experiments are some of these. In this case, the availability of a tunnel dug inside a mountain to shelter the experiment from cosmic rays is invaluable. Similar experiments are performed underground in deep mines. Years of experimental efforts have not turned out exotic dark matter particles yet, and several scientists are now contemplating alternative possibilities, but the search goes on.

Chapter 8

Epilogue

We have discussed several phenomena that are seen in the alpine environment. The science involved spans mechanics, fluid mechanics, thermal physics, gravity, electromagnetism, non-linear dynamics, atmospheric physics, geology, glaciology, chemistry, theoretical biology, and we have pointed out also connections with relativity, quantum mechanics, statistical mechanics, particle physics, astrophysics, cosmology, and astronomy. Although we have tried to separate these areas of science and the various aspects that belong to them, often many of these aspects occur simultaneously and their separation then becomes artificial. This complication happens because many things are interconnected in nature, a characteristic encountered in most problems of environmental physics [Boeker and van Grondelle (2011); Guyot (1998); Campbell (1977); Campbell and Norman (1998); Henry and Heinke (1989); Faraoni (2006)]. As John Muir said, "When we try to pick out anything by itself, we find it hitched to everything else in the Universe" [Muir (1911)].

Very little detail has been provided here. Indeed, any of the physics problems touched upon would require a much longer discussion and sometimes a sophisticated approach with analytical, numerical, and statistical methods. This level of treatment would require extensive knowledge of many specialized fields which is usually not found in individual researchers. The high specialization of modern science would require a large team of scientists and several coordinators to address all these problems satisfactorily. It is a privilege of mountain climbers to stumble on, and be aware of, all these issues in a single day when wandering in the hills and being immersed in these phenomena. In this book I did not offer extensive knowledge nor understanding, indeed I just whetted the appetite providing no deep nor detailed answers to the many questions that I raised and to those that were not spelled out, but follow naturally. Some of these questions do not yet have a partial or full answer. The bibliography should be useful to those willing to find out more. This book will have reached its goal if it stimulates its readers to reason and to find out why things are the way they are and happen the way they do. Asking the simple questions "why" and "how", when looking at a rock marked by ancient glaciers or when taking in a vast panorama raises many complex issues, some of which have been discussed at least partially. Understanding the natural phenomena around us, which are more dramatic in the mountains, makes us a bit more part of nature rather than entities separated from it and maybe this is one of the implicit goals of mountain climbing. Overall, it is impossible to describe mountaineering, it must be experienced. Similarly, it is almost impossible for people who live in cities in flat geographical areas to understand mountain culture, which has some universal traits but also a very distinctive regional character. Perhaps science can do a bit better in bringing people closer to the mountain environment through reasoning. "Intellectual climbing" is not a substitute for physical climbing, but there is little doubt that there is a place for it somewhere in the life of a mountaineer. Non-climbers who participate only in this intellectual climbing miss out on the best part, but at least they get something out of the mountains, too, and that's better than nothing, and perhaps that's all they care about. Climbers will add some extra understanding to that body of knowledge, often subconscious, that is accumulated over years of climbing. Happy trails!

Appendix A

Here we provide the values of physical constants used in the text and the expressions of common linear differential operators.

A.1 Physical constants

A.1.1 *Fundamental constants*

Gravitational constant	$G = 6.673 \cdot 10^{-11} \, \mathrm{N} \cdot \mathrm{m}^2 \cdot \mathrm{kg}^{-2}$
speed of light *in vacuo*	$c = 2.998 \cdot 10^8 \, \mathrm{m/s}$
Boltzmann constant	$K = 1.381 \cdot 10^{-23} \, \mathrm{J/K}$
Stefan–Boltzmann constant	$\sigma = 5.671 \cdot 10^{-8} \, \mathrm{W} \cdot \mathrm{m}^{-2} \cdot \mathrm{K}^{-4}$
universal gas constant	$R = 8.314 \, \mathrm{J} \cdot \mathrm{K}^{-1} \cdot \mathrm{mol}^{-1}$
Avogadro's number	$N_A = 6.022 \cdot 10^{23} \, \mathrm{mol}^{-1}$

A.1.2 *Astronomical constants*

mass of the Earth	$M_E = 5.978 \cdot 10^{24} \, \mathrm{kg}$
mass of the Sun	$M_\odot = 1.989 \cdot 10^{30} \, \mathrm{kg}$
mass of the Moon	$M_M = 7.35 \cdot 10^{22} \, \mathrm{kg}$
average radius of the Earth	$R_E = 6.370 \cdot 10^6 \, \mathrm{m}$
equatorial radius of the Earth	$R = 6.378 \cdot 10^6 \, \mathrm{m}$
radius of the Sun	$R_\odot = 6.96 \cdot 10^8 \, \mathrm{m}$
radius of the Moon	$R_M = 1.74 \cdot 10^6 \, \mathrm{m}$
average Sun-Earth distance	$1 \, \mathrm{A.U.} = 1.496 \cdot 10^{11} \, \mathrm{m}$

solar constant $S = 1.370 \cdot 10^3 \text{ W/m}^2$
average acceleration of gravity $g = 9.81 \text{ m/s}^2$
on the Earth's surface

A.1.3   *Microscopic physics*

electron charge $e = 1.602 \cdot 10^{-19} \text{ C}$
electron mass $m_e = 9.109 \cdot 10^{-31} \text{ kg} = 511.0 \text{ keV}$
proton mass $m_p = 1.673 \cdot 10^{-27} \text{ kg} = 938.3 \text{ MeV}$
atomic mass unit $1 \text{ a.m.u.} = 1.661 \cdot 10^{-27} \text{ kg} = 931.5 \text{ MeV}$
Planck constant $h = 6.625 \cdot 10^{-34} \text{ J} \cdot \text{s}$
reduced Planck constant $\hbar = \frac{h}{2\pi} = 1.0545 \cdot 10^{-34} \text{ J} \cdot \text{s}$

A.1.4   *Air*

density of air at 10°C $\rho = 1.247 \text{ kg/m}^3$
density of air at 20°C $\rho = 1.205 \text{ kg/m}^3$
specific heat of dry air $c_p = 1004 \text{ J} \cdot \text{K}^{-1} \cdot \text{kg}^{-1}$
 at constant pressure
thermal conductivity at 27°C $k = 0.026 \text{ W} \cdot \text{m}^{-1} \cdot (°\text{C})^{-1}$
speed of sound at 1 atm and 0°C $v_s = 330 \text{ m/s}$

A.1.5   *Water*

density of freshwater at 20°C $\rho = 0.998 \cdot 10^3 \text{ kg/m}^3$
latent heat of fusion $3.34 \cdot 10^5 \text{ J/kg} = 79.7 \text{ cal/g}$
latent heat of vaporization $2.26 \cdot 10^6 \text{ J/kg} = 539 \text{ cal/g}$
coefficient of thermal $2.10 \cdot 10^{-4} \ (°\text{C})^{-1}$
 expansion at 20°C
specific heat at 25°C $4186 \text{ J} \cdot \text{kg}^{-1} \cdot (°\text{C})^{-1} = 1.00 \text{ cal/g}$
surface tension at 20°C $7.28 \cdot 10^{-2} \text{ N/m}$

A.1.6   *Conversion factors*

$1 \text{ kWh} = 3.6 \cdot 10^6 \text{ J}$
$1 \text{ atm} = 1.01325 \cdot 10^5 \text{ Pa}$
$1 \text{ Å} = 10^{-10} \text{ m}$

$1\,\mathrm{fm} = 10^{-15}\,\mathrm{m}$
$1\,\mathrm{eV} = 1.60 \cdot 10^{-19}\,\mathrm{J}$

A.2 Differential operators

Here we report the expressions of the most common linear differential operators in three spatial dimensions, in the three coordinate systems most widely used (*i.e.*, Cartesian, cylindrical, and spherical coordinates). The triads $\{\vec{e}_x, \vec{e}_y, \vec{e}_z\}$, $\{\vec{e}_r, \vec{e}_\varphi, \vec{e}_z\}$, and $\{\vec{e}_r, \vec{e}_\theta, \vec{e}_\varphi\}$ are the unit vectors along the fundamental directions of the respective coordinate system.

A.2.1 *Gradient*

The gradient of a function f in Cartesian coordinates (x, y, z) is

$$\vec{\nabla} f = \frac{\partial f}{\partial x}\,\vec{e}_x + \frac{\partial f}{\partial y}\,\vec{e}_y + \frac{\partial f}{\partial z}\,\vec{e}_z.$$

The gradient of f in cylindrical coordinates (r, φ, z) is

$$\vec{\nabla} f = \frac{\partial f}{\partial r}\,\vec{e}_r + \frac{1}{r}\frac{\partial f}{\partial \varphi}\,\vec{e}_\varphi + \frac{\partial f}{\partial z}\,\vec{e}_z.$$

The gradient of f in spherical polar coordinates (r, θ, φ) is

$$\vec{\nabla} f = \frac{\partial f}{\partial r}\,\vec{e}_r + \frac{1}{r}\frac{\partial f}{\partial \theta}\,\vec{e}_\theta + \frac{1}{r \sin \theta}\frac{\partial f}{\partial \varphi}\,\vec{e}_\varphi.$$

A.2.2 *Divergence*

The divergence of a vector field $\vec{a}$ in Cartesian coordinates (x, y, z) is

$$\vec{\nabla} \cdot \vec{a} = \frac{\partial a^x}{\partial x} + \frac{\partial a^y}{\partial y} + \frac{\partial a^z}{\partial z}.$$

The divergence of $\vec{a}$ in cylindrical coordinates (r, φ, z) is

$$\vec{\nabla} \cdot \vec{a} = \frac{1}{r}\frac{\partial}{\partial r}(r a_r) + \frac{1}{r}\frac{\partial a_\theta}{\partial \theta} + \frac{\partial a_z}{\partial z}.$$

The divergence of $\vec{a}$ in spherical polar coordinates (r, θ, φ) is

$$\vec{\nabla} \cdot \vec{a} = \frac{1}{r^2} \frac{\partial}{\partial r} \left(r^2 a_r\right) + \frac{1}{r \sin \theta} \frac{\partial}{\partial \theta} \left(a_\theta \sin \theta\right) + \frac{1}{r \sin \theta} \frac{\partial a_\varphi}{\partial \varphi}.$$

A.2.3   *Laplacian*

The Laplacian of a function f in Cartesian coordinates (x, y, z) is

$$\nabla^2 f = \frac{\partial^2 f}{\partial x^2} + \frac{\partial^2 f}{\partial y^2} + \frac{\partial^2 f}{\partial z^2}.$$

The Laplacian of f in cylindrical coordinates (r, φ, z) is

$$\nabla^2 f = \frac{1}{r} \frac{\partial}{\partial r} \left(r \frac{\partial f}{\partial r}\right) + \frac{1}{r^2} \frac{\partial^2 f}{\partial \varphi^2} + \frac{\partial^2 f}{\partial z^2}.$$

The Laplacian of f in spherical polar coordinates (r, θ, φ) is

$$\nabla^2 f = \frac{1}{r^2} \frac{\partial}{\partial r} \left(r^2 \frac{\partial f}{\partial r}\right) + \frac{1}{r^2 \sin \theta} \frac{\partial}{\partial \theta} \left(\sin \theta \frac{\partial f}{\partial \theta}\right) + \frac{1}{r^2 \sin^2 \theta} \frac{\partial^2 f}{\partial \varphi^2}.$$

Appendix B

Here we derive the equation of hydrostatic equilibrium in its differential form. Let the z-axis point vertically upward and consider an elementary fluid parcel with the shape of a vertical parallelepiped of volume δV. This parallelepiped has horizontal faces of area A at levels z and $z + \delta z$ (Fig. 5.3). The forces acting in the vertical direction are the weight $-(\delta m)g$ of the parcel (pointing downward), where g is the acceleration of gravity, and those caused by the pressure on the horizontal faces, $-AP(z + \delta z)$, directed downward, and $AP(z)$ directed upward. The mass of the parcel is $\delta m = \rho \delta V = \rho A \delta z$, where ρ is the density of the fluid. In equilibrium, the pressure gradient balances the weight. The balance in the vertical direction reads

$$A\left[P(z) - P\left(z + \delta z\right)\right] - \rho g A \delta z = 0.$$

Re-arranging, we have

$$\frac{P\left(z + \delta z\right) - P(z)}{\delta z} = -\rho g$$

and, by taking the limit $\delta z \to 0$, we obtain the equation of hydrostatic equilibrium

$$\frac{dP}{dz} = -\rho g$$

or, using differentials,

$$dP = -\rho g \, dz.$$

The forces acting on the fluid parcel in the horizontal direction, and applied to the parcel sides, are equal and opposite and balance each other.

Appendix C

Here we expand a little the discussion of the classic catenary problem and show how the catenary curve relates to a variational principle. Consider a heavy string hanging from two fixed points in a vertical (x, y) plane and described by the profile $y(x)$. The linear density of the string is $\mu = dm/ds$, where $ds = \sqrt{dx^2 + dy^2} = \sqrt{1 + (y')^2}\,dx$ is the line element along the string and where $y' \equiv dy/dx$. The gravitational potential energy of an element of string of length ds located at horizontal position x is $dE_g = \mu g y(x) ds$. The total gravitational potential energy of a string suspended by two points of horizontal coordinates x_1 and x_2 is the functional of the curve $y(x)$

$$E_g\left[y(x)\right] = \mu g \int_{x_1}^{x_2} dx\, y \sqrt{1 + (y')^2} \equiv \int_{x_1}^{x_2} L\, dx.$$

The Lagrangian $L\left(y(x), y'(x)\right)$ does not depend explicitly on the coordinate x and, therefore, the corresponding Hamiltonian function is conserved [Goldstein (1980)]:

$$\mathcal{H} = \frac{\partial L}{\partial(y')}\, y' - L = c_1,$$

where c_1 is a constant. This is a classic result of mechanics [Goldstein (1980)], which is easy to verify directly. This equation amounts to writing

$$\frac{-y}{\sqrt{1 + (y')^2}} = c_1.$$

In order to solve this equation, we manipulate and rewrite it as

$$\frac{y'}{\sqrt{(y/c_1)^2 - 1}} = \pm 1$$

and, integrating with respect to x,

$$c_1 \int dx \, \frac{y'}{\sqrt{y^2 - c_1^2}} = \pm \int dx,$$

or

$$c_1 \int \frac{dy}{\sqrt{y^2 - c_1^2}} = \pm x + \text{const.}$$

The integral on the left hand side is computed, giving

$$\ln\left(y + \sqrt{y^2 - c_1^2}\right) = \pm \frac{x}{c_1} + \text{const.}$$

Exponentiating both sides produces

$$y + \sqrt{y^2 - c_1^2} = D \exp\left(\pm \frac{x}{c_1}\right),$$

where D is an integration constant. By isolating the square root and squaring both sides of the resulting equation, two y^2 terms cancel out and one is left with

$$y = \frac{D e^{\pm x/c_1} + \frac{c_1^2}{D} e^{\mp x/c_1}}{2}.$$

The choice $D = c_1$ then provides the catenary solutions

$$y(x) = c_1 \cosh\left(\frac{x}{c_1}\right).$$

Appendix D

Here we prove that

$$s_{xx}^2 + s_{yy}^2 + s_{zz}^2 + 2\left(s_{xy}^2 + s_{yz}^2 + s_{xz}^2\right)$$
$$= 2\left(-s_{xx}s_{yy} - s_{xx}s_{zz} - s_{yy}s_{zz} + s_{xy}^2 + s_{yz}^2 + s_{xz}^2\right), \qquad \text{(D.1)}$$

which is Eq. (3.28). We have

$$s_{xx}^2 + s_{yy}^2 + s_{zz}^2 + 2\left(s_{xy}^2 + s_{yz}^2 + s_{xz}^2\right)$$
$$= s_{xx}^2 + s_{yy}^2 + s_{zz}^2 + 2\left(s_{xx}s_{yy} + s_{xx}s_{zz} + s_{yy}s_{zz}\right)$$
$$\quad - 2\left(s_{xx}s_{yy} + s_{xx}s_{zz} + s_{yy}s_{zz}\right) + 2\left(s_{xy}^2 + s_{yz}^2 + s_{xz}^2\right)$$
$$= \left(s_{xx} + s_{yy} + s_{zz}\right)^2 + 2\left(-s_{xx}s_{yy} - s_{xx}s_{zz}\right.$$
$$\left. - s_{yy}s_{zz} + s_{xy}^2 + s_{yz}^2 + s_{xz}^2\right) \qquad \text{(D.2)}$$

where, in the last line, we used the fact that $\mathrm{Tr}(\hat{\mathbf{s}}) = 0$.

Bibliography

O. Abe, Shear strength and angle of repose of snow layers including graupel, *Annals of Glaciology* **38**, 305–308 (2004).

I.J.R. Aitchison, *Supersymmetry in Particle Physics*. Cambridge University Press, Cambridge, UK (2007).

R. Alley, Firn densification by grain-boundary sliding: a first model, *Journal de Physique Colloques* **48**, 249–256 (1987).

ALMA Observatory website http://www.almaobservatory.org/en/home/

L. Amendola and S. Tsujikawa, *Dark Energy, Theory and Observations*. Cambridge University Press, Cambridge, UK (2010).

K.N. Ananda and M. Bruni, Cosmological dynamics and dark energy with non-linear equation of state: a quadratic model, *Physical Review D* **74**, 023523 (2006).

K.N. Ananda and M. Bruni, Cosmological dynamics and dark energy with a quadratic equation of state: anisotropic models, large-scale perturbations and cosmological singularities, *Physical Review D* **74**, 023524 (2006).

D.L. Anderson and C.S. Benson, The densification and diagenesis of snow, in *Ice and Snow: Properties, Processes and Applications*, Proceedings, MIT February 12–16, 1962, W.D. Kingery *et al.* eds., pp. 391–411. MIT Press, Cambridge, Mass., USA (1962).

P.S. Aplin and D.J. Hill, Growth analysis of circular lichen thalli, *Journal of Theoretical Biology* **78**, 347–363 (1979).

R.J. Arthern, H.F.J. Corr, F. Gillet-Chaulet, R.L. Hawley, and E.M. Morris, Inversion of the density-depth profile of polar firn using a stepped-frequency radar, *Journal of Geophysical Research: Earth and Surface Processes* **118**, 1257–1263 (2013).

Atacama Pathfinder EXperiment (APEX) website http://www.apex-telescope.org/

H. Bader, Sorge's law of densification of snow on high polar glaciers, *Journal of Glaciology* **2**, 319–323 (1954).

H. Baer and X. Tata, *Weak Scale Supersymmetry*. Cambridge University Press, Cambridge, UK (2006).

J. Bailey, K. Borer, F. Combley, H. Drumm, F. Krienen, F. Lange, E. Picasso, W. von Ruden, F.J.M. Farley, J.H. Field, W. Flegel, and P.M. Hattersley, Measurements of relativistic time dilatation for positive and negative muons in a circular orbit, *Nature* **268**, 301–305 (1977).

B. Barris *et al.*, Twenty-three high-redshift supernovae from the Institute for Astronomy Deep Survey: doubling the supernova sample at $z > 0.7$, *Astrophysical Journal* **602**, 571–594 (2004).

J.D. Barrow, Sudden future singularities, *Classical and Quantum Gravity* **21**, L79–L82 (2004).

J.D. Barrow, *Mathletics*. W.W. Norton & C., New York (2012).

M.R. Bennett and N.F. Glasser, *Glacial Geology, Ice Sheets and Landforms*, second edition. Wiley-Blackwell, Chichester, UK (2009).

C.S. Benson, *Stratigraphic studies in the snow and firn of the Greenland ice sheet*, US Snow, Ice and Permafrost Research Establishment, Research Report 70 (1962).

V. Bergeron, C. Berger, and M.D. Betterton, Controlled irradiative formation of penitentes, *Physical Review Letters* **96**, 098502 (2006).

C. Bergmann, Über die Verhältnisse der Wärmeökonomie der Thiere zu ihrer Grösse, *Göttinger Studien* **3** (1), 595–708 (1847).

G. Bertone, D. Hooper, and J. Silk, Particle dark matter: evidence, candidates and constraints, *Physics Reports* **405**, 279–390 (2005).

M.D. Betterton, Theory of structure formation in snowfields motivated by penitentes, suncups, and dirt cones, *Physical Review E* **63**, 056129 (2001).

P. Binétruy, *Supersymmetry: Theory, Experiment, and Cosmology*. Oxford University Press, Oxford, UK (2006).

J. Binney and S. Tremaine, *Galactic Dynamics*. Princeton University Press, Princeton, NJ (1987).

E. Boeker and R. van Grondelle, *Environmental Physics*. J. Wiley & Sons, Chicester, UK (2011).

A. Borowiec, M. Kamionka, A. Kurek, and M. Szydlowski, Cosmic acceleration from modified gravity with Palatini formalism, *Journal of Cosmology and Astroparticle Physics* **2012**, 27 (2012).

A. Borowiec, A. Stachowski, M. Szydlowski, and A. Wojnar, Inflationary cosmology with Chaplygin gas in Palatini formalism, *Journal of Cosmology and Astroparticle Physics* **2016** (01), 40 (2016).

M. Bouhmadi-López, A. Errahmani, P. Martin-Moruno, T. Ouali, and Y. Tavakoli, The little sibling of the big rip singularity, *International Journal of Modern Physics D* **24**, 1550078/1–20 (2015).

F. Brauer and J.A. Noel, *Introduction to Differential Equations With Applications*. Harper and Row, New York (1986).

J.F. Campbell, *Frost and Fire*. J.B. Lippincott, Philadelphia (1865).

I.M. Campbell, *Energy and the Atmosphere-A Physical-Chemical Approach*. J. Wiley & Sons, London (1977).

G.S. Campbell and J.M. Norman, *An Introduction to Environmental Biophysics*, second edition. Springer-Verlag, New York (1998).

M.E. Caplan, Calculating the potato radius of asteroids using the height of Mt. Everest, preprint arXiv:1511.04297 (2015).

S. Capozziello, V.F. Cardone, E. Elizalde, S. Nojiri, and S.D. Odintsov, Observational constraints on dark energy with generalized equations of state, *Physical Review D* **73**, 043512 (2006).

S.M. Carroll, *Spacetime and Geometry: An Introduction to General Relativity.* Addison Wesley, San Francisco, USA (2004).

A.H. Carter, *Classical and Statistical Thermodynamics.* Prentice Hall, Upper Saddle River, NJ (2001).

CERN home page https://home.cern

Cerro Armazones Observatory website https://www.eso.org/public/images/armaz-bochum-pan/

Cerro Tololo Inter-American Observatory website http://www.ctio.noao.edu/noao/

A. Chen, G.W. Gibbons, and Y. Yang, Explicit integration of Friedmann's equation with nonlinear equations of state, *Journal of Cosmology and Astroparticle Physics* **05**, 020/1–43 (2015).

A. Chen, G.W. Gibbons, and Y. Yang, Friedmann-Lemaitre cosmologies via roulettes and other analytic methods, *Journal of Cosmology and Astroparticle Physics* **05**, 056 (2015).

S. Childress and J.B. Keller, Lichen growth, *Journal of Theoretical Biology* **32**, 157–165 (1980).

C. Clauser, *Thermal storage and transport properties of rocks, I: heat capacity and latent heat, Encyclopedia of Solid Earth Geophysics*, H.K. Gupta ed., pp. 1423–1430. Springer, Dordrecht (2011).

W. Colgan, H. Rajaram, W. Abdalati, C. McCutchan, R. Mottram, M.S. Moussavi, and S. Grigsby, Glacier crevasses: observations, models, and mass balance implications, *Reviews of Geophysics* **54**, 119–161 (2016).

W.N. Cottingham and D.A. Greenwood, *An Introduction to the Standard Model of Particle Physics.* Cambridge University Press, Cambridge, UK (2003).

K.M. Cuffey and W.S.B. Paterson, *The Physics of Glaciers.* Elsevier, Amsterdam (2010).

T. Daffern, *Avalanche Safety for Skiers and Climbers.* Rocky Mountain Books, Calgary (1992).

C. Darwin, *The Voyage of the Beagle.* Penguin, London, UK (1989).

W. Davis, *Into the Silence: The Great War, Mallory, and The Conquest of Everest.* Vintage, Toronto (2012).

N. Dragon, *The Geometry of Special Relativity — A Concise Course.* Springer, New York (2012).

R. Durrer, *The Cosmic Microwave Background.* Cambridge University Press, Cambridge, UK (2008).

P.-G. de Gennes, F. Brochard-Wyart, and D. Quéré, *Capillarity and Wetting Phenomena.* Springer, New York (2010).

N. Easwar and D.A. MacIntire, Study of the effect of relativistic time dilation on cosmic ray muon flux. An undergraduate modern physics experiment, *American Journal of Physics* **59**, 589–592 (1991).

A. Einstein, Über die von der molekularkinetischen Theorie der Wärme geforderte Bewegung von in ruhenden Flüssigkeiten suspendierten Teilchen, *Annalen der Physik* **17**, 549–560 (1905).

G. Ekstrom, M. Nettles, and V.C. Tsai, Seasonality and increasing frequency of Greenland glacial earthquakes, *Science* **311**, 1756–1757 (2006).

L.J. Elms, *Beyond Nootka, A Historical Perspective of Vancouver Island Mountains*. Misthorn Press, Courtenay, BC, Canada (1996).

Extremely Large Telescope (ELT) website https://www.eso.org/public/telesinstr/elt/

J. Failletaz, M. Funk, and D. Sornette, Icequakes as precursors of ice avalanches, preprint arXiv:0906.5528.

J. Failletaz, M. Funk, and D. Sornette, *Prediction of alpine glacier sliding and instabilities: a new hope*, Proceedings, EGU General Assembly Vienna, Austria 22–27 April 2012, p. 7243 [preprint arXiv:1201.1189 (2012)].

J. Failletaz, D. Sornette, and M. Funk, Climate warming and stability of cold hanging glaciers: lessons from the gigantic 1895 Altels break-off, preprint arXiv:1101.5062 (2011).

V. Faraoni, *Exercices in Environmental Physics*. Springer, New York (2006).

V. Faraoni, *Special Relativity*. Springer, New York (2013).

V. Faraoni, Glacier physics from first principles, in *Glaciers: Formation, Climate Change and their Effects*, N. Doyle ed. Nova Science, New York (2016).

V. Faraoni and A.M. Cardini, Analogues of glacial valley profiles in particle mechanics and in cosmology, *FACETS* **2**, 286300 (2017).

V. Faraoni and M.W. Vokey, The thickness of glaciers, *European Journal of Physics* **6**, 055031 (2015).

M. Fayngold, *Special Relativity and How It Works*. Wiley, New York (2008).

E. Fermi, *Thermodynamics*. Dover, New York (1956).

G.L. Fowles and G.L. Cassiday, *Analytical Mechanics*. Thomson/Brooks-Cole, Belmont, CA (2005).

P.H. Frampton, K.J. Ludwick, and R.J. Scherrer, The little rip, *Physical Review D* **84**, 063003/1–5 (2011).

J.A. Fredston and D. Fesler, *Snow Sense: A Guide to Evaluating Snow Avalanche Hazard*. Alaska Mountain Safety Center, Anchorage (1988).

S. Gasiorowicz, *Quantum Physics*, third edition. Wiley, New York (2003).

Gemini Telescope website http://www.gemini.edu/

J. Gleick, *Chaos, Making A New Science*, second edition. Penguin, New York (2008).

J.W. Glen, Experiments on the deformation of ice, *Journal of Glaciology* **2**, 111–114 (1952).

J.W. Glen, The creep of polycrystalline ice, *Proceedings of the Royal Society of London A* **228**, 519–538 (1955).

D. Goddard and U. Neumann, *Performance Rock Climbing*. Stackpole Books, Mechanicsbsurg, PA (1993).

H. Goldstein, *Classical Mechanics*. Addison-Wesley, Reading, MA (1980).

M. Górski, *Seismic Events in Glaciers*. Springer, New York (2014).

R. Greve and H. Blatter, *Dynamics of Ice Sheets and Glaciers.* Springer, New York (2009).

D.J. Griffiths, *Introduction to Electrodynamics,* third edition. Prentice Hall, Upper Saddle River, NJ (1999).

D.J. Griffiths, *Introduction to Quantum Mechanics.* Prentice Hall, Upper Saddle River, NJ (2005).

D.J. Griffiths, *Introduction to Elementary Particles,* second edition. Wiley, New York (2008).

P. Grohmann, *Wanderungen in den Dolomiten.* Verlag von Carl Gerhold's Sohn, Vienna (1877).

J. Guckenheimer and P. Holmes, *Nonlinear Oscillations, Dynamical Systems and Bifurcation of Vector Fields.* Springer, New York (1983).

G. Guyot, *Physics of the Environment and Climate.* J. Wiley & Sons/Praxis Publishing, Chicester, UK (1998).

D. Halliday, R. Resnick, and J. Walker, *Fundamentals of Physics,* seventh edition. Wiley, New York (2005).

J. Harte, *Consider a Spherical Cow: A Course in Environmental Problem Solving.* University Science Books, Sausalito, CA (1988).

J. Harte, *Consider a Cylindrical Cow: More Adventures in Environmental Problem Solving.* University Science Books, Sausalito, CA (2001).

J.B. Hartle, *Gravity, An Introduction to Einstein's General Relativity.* Addison Wesley, San Francisco, USA (2003).

R.L. Hawley, E.M. Morris, and J.R. McConnell, Rapid techniques for determining annual accumulation applied at Summit, Greenland, *Journal of Glaciology* **54**, 838–845 (2008).

J.G. Henry and G.W. Heinke, *Environmental Science and Engineering.* Prentice Hall, Englewood Cliffs, NJ (1989).

D.J. Hill, The growth of lichens with special references to the modeling of circular thalli, *Lichenologist* **13**, 265–287 (1981).

E. Hille, *Lectures on Ordinary Differential Equations.* Addison-Wesley, Reading, MA (1969).

M. Hirano and M. Aniya, A rational explanation of cross-profile morphology for glacial valleys and of glacial valley development, *Earth Surface Processes and Landforms* **13**, 707–716 (1988).

R. LeB. Hooke, *Principles of Glacier Mechanics,* second edition. Cambridge University Press, Cambridge, UK (2005).

J.D. Jackson, *Classical Electrodynamics,* second edition. J. Wiley & Sons, New York (1975).

R. Jayawardhana, *Strange New Worlds.* Harper-Collins, Toronto (2011).

R. Knop *et al.,* New Constraints on Ω_M, Ω_Λ, and w from an independent set of 11 high-redshift supernovae observed with the Hubble Space Telescope, *Astrophysical Journal* **598**, 102–137 (2003).

E.P. Krider, Benjamin Franklin and lightning rods, *Physics Today* **59**, 1, 42–48 (2006).

P. Labelle, *Supersymmetry Demystified.* McGraw-Hill, New York (2010).

L.D. Landau and E.M. Lifshitz, *The Classical Theory of Fields*, fourth reprinted English edition. Pergamon Press, Oxford, UK (1989).

G.A. Landis, Physics of mountains, *American Journal of Physics* **54**, 871 (1986).

S. Lang, *Linear Algebra*, third edition. Springer, New York (1987).

Large Synoptic Survey Telescope (LSST) website https://www.lsst.org/

B. Lawn, *Fracture of Brittle Solids*. Cambridge University Press, Cambridge, UK (1993).

Las Campanas Observatory website http://www.lco.cl/

La Silla Observatory website https://www.eso.org/public/teles-instr/lasilla/

D.J. Levitin, *The Organized Mind: Thinking Straight in The Age of Information Overload*. Dutton/Penguin Random House, New York (2014).

K. Libbrecht and P. Rasmussen, *The Snowflake, Winter's Secret Beauty*. Voyager Press/MBI Publishing, St. Paul, MN, USA (2003).

A. Liddle, *An Introduction to Modern Cosmology*. Wiley, Chichester, UK (2003).

C.H. Lineweaver and M. Norman, The potato radius: a lower minimum size for dwarf planets, *Australian Space Science Conference Series: Proceedings of the 9th Australian Space Science Conference*, 67–78 (2009).

J. Long, *Climbing Anchors*. Chockstone Press, Evergreen, CO, USA (1993).

E.J. Lopez Garcia, E.G. Rivera-Valentin, P.M. Schenk, N.P. Hammonda, and A.C. Barr, Topographic constraints on the origin of the equatorial ridge on Iapetus, *Icarus* **237**, 419–421 (2014).

F.K. Lutgens and E.J. Tarbuck, *Essentials of Geology*, seventh edition. Prentice Hall, Upper Saddle River, NJ (2000).

I.G. Main, *Vibrations and Waves in Physics*, third edition. Cambridge University Press, Cambridge, UK (1993).

N.J. Mason and P. Hughes, *Introduction to Environmental Physics: Planet Earth, Life, and Climates*. Taylor & Francis, New York (2001).

W.J. McGee, Glacial canons, *Journal of Geology* **2**, 350–364 (1894).

A. Messiah, *Quantum Mechanics*. North-Holland, Amsterdam (1961).

C.W. Misner, K.S. Thorne, and J.A. Wheeler, *Gravitation*. Freeman, New York (1973).

B.L. Moiseiwitsch, *Variational Principles*. Interscience, London, UK (1966).

J.L. Monteith and M.H. Unsworth, *Principles of Environmental Physics*. Butterworth-Heinemann, London, UK (1990).

Great Britain Meteorological Office, *Symons' Monthly Meteorological Magazine* **18**, 81–82 (1883).

F. Morgan, A note on cross-profile morphology for glacial valleys, *Earth Surface Processes and Landforms* **30**, 513–514 (2005).

J. Muir, *My First Summer in the Sierra*. Houghton and Mifflin, Boston (1911).

E. Newby, *Great Ascents, A Narrative History of Mountaineering*. David & Charles, Newton Abbot, UK (1977).

S. Nojiri and S.D. Odintsov, Final state and thermodynamics of a dark energy universe, *Physical Review D* **70**, 103522 (2004).

S. Nojiri and S.D. Odintsov, Inhomogeneous equation of state of the universe: phantom era, future singularity, and crossing the phantom barrier, *Physical Review D* **72**, 023003 (2005).

J.F. Nye, The mechanics of glacier flow, *Journal of Glaciology* **2**, 82–93 (1952).

H.J. Pain, *The Physics of Vibrations and Waves*. Wiley, New York (2005).

Paranal Observatory website https://www.eso.org/public/teles-instr/paranal-observatory/

A.F. Parker-Rhodes, Fairy ring kinetics, *Transactions of the British Mycological Society* **38**, 59–72 (1955).

Particle Data Group website http://pdg.lbl.gov/

W.S.B. Paterson, *The Physics of Glaciers*, third edition. Butterworth-Heinemann, Oxford, UK (1994).

F. Pattyn and W. Van Huele, Power law or power flaw?, *Earth Surface Processes and Landforms* **23**, 761–767 (1998).

S. Perlmutter *et al.*, Discovery of a supernova explosion at half the age of the universe, *Nature* **391**, 51–54 (1998).

S. Perlmutter *et al.*, Measurements of Ω and Λ from 42 high-redshift supernovae, *Astrophysical Journal* **517**, 565–586 (1999).

M.F. Perutz, Glaciology — the flow of glaciers, *The Observatory* **70**, 64–65 (1950).

I. Peterson, *A Mathematical Mistery Cruise*. Freeman & C., New York (1990), pp. 111–150.

G. Pole, *The Canadian Rockies, A History in Photographs*. Altitude Publishing, Canmore, Canada (1991).

G. Pole, *Summit Tales, Early Adventures in the Canadian Rockies*. Altitude Publishing, Canmore, Canada (2005).

A. Post and E.R. LaChapelle, *Glacier Ice*, revised edition. University of Washington Press, Seattle (2000).

P. Preuss, quoted in R. Messner, *Freiklettern Mit Paul Preuss*. BLV Verlaggesellschaft, Munich (1986).

M.C.F. Proctor, The growth curve of the crustose lichen Buellia Canescens (Dicks.) de Not., *New Phytologist* **79**, 659–663 (1977).

J. Rafelski, *Relativity Matters*. Springer, New York (2017).

V.A. Rakov and M.A Uman, *Lightning: Physics and Effects*. Cambridge University Press, Cambridge, UK (2007).

J.R. Reitz and F.J. Milford, *Foundations of Electromagnetic Theory*. Addison-Wesley, Reading, MA (1960).

M.J. Reidy, John Tyndall's vertical physics: from rock quarries to icy peaks, *Physics in Perspective* **12**, 122–145 (2010).

H.-G. Richardi, *La Conquista delle Dolomiti*, Athesia, Bolzano, Italy (2008).

A.G. Riess *et al.*, Observational evidence from supernovae for an accelerating universe and a cosmological constant, *Astronomical Journal* **116**, 1009–1038 (1998).

A.G. Riess, A.V. Filippenko, W. Li, and B.P. Schmidt, Is there an indication of evolution of type Ia supernovae from their rise times?, *Astronomical Journal* **118**, 2668–2674 (1999).

A.G. Riess *et al.*, The farthest known supernova: support for an accelerating universe and a glimpse of the epoch of deceleration, *Astrophysical Journal* **560**, 49–71 (2001).

A.G. Riess *et al.*, Type Ia supernova discoveries at $z > 1$ from the Hubble Space Telescope: evidence for past deceleration and constraints on dark energy evolution, *Astrophysical Journal* **607**, 665–687 (2004).

W. Rindler, *Introduction to Special Relativity*, second edition. Clarendon Press, Oxford, UK (1991).

R. Robbins, *Basic Rockraft*. La Siesta Press, Glendale, CA (1971).

R. Robbins, *Advanced Rockraft*. La Siesta Press, Glendale, CA (1977).

B. Rossi and D.B. Hall, Variation of the rate of decay of mesotrons with momentum, *Physical Review* **59**, 223–228 (1941).

P. Schenk, D.P. Hamilton, R.E. Johnson, W.B. McKinnon, C. Paranicas, J. Schmidt, and M.R. Showalter, Plasma, plumes and rings: Saturn system dynamics as recorded in global color patterns on its midsize icy satellites, *Icarus* **211**, 740–757 (2011).

P.A.G. Scheuer, How high can a mountain be?, *Journal of Astrophysics and Astronomy* **2**, 165–169 (1981).

L.I. Schiff, *Quantum Mechanics*, third edition. McGraw-Hill, New York (1968).

P. Schneider, J. Ehlers, and E.E. Falco, *Gravitational Lenses*. Springer-Verlag, New York (1992).

A.C. Scott (ed.), *Enciclopedia of Nonlinear Science*, p. 291. Routledge, New York (2005).

A.C. Scott, *The Nonlinear Universe: Chaos, Emergence, Life*. Springer, Berlin (2007).

V. Schytt, *Glaciology II. Snow studies at Maudheim — Snow studies inland — The inner structure of the ice shelf of Maudheim as shown by core drilling*, Norwegian-British-Swedish Antarctic expedition 1949–52. Scientific Results vol. 4: A, B, C (1958).

F.W. Sears and G.L. Salinger, *Thermodynamics, Kinetic Theory, and Statistical Thermodynamics*. Addison-Wesley, Reading, MA (1975).

H. Seddik, R. Greve, and S. Sugiyama, Numerical simulation of the evolution of glacial valley cross sections, preprint arXiv:0901.1177 (2009).

M. Segre, Galileo, Viviani, and the tower of Pisa, *Studies in History and Philosophy of Science Part A* **20**(4), 435–451 (1989).

S. Silva e Costa, An entirely analytical cosmological model, *Modern Physics Letters A* **24**, 531–540 (2009).

J. Simpson, *Touching the Void*. Pan Books, London, UK (1989).

I.M. Singer and J.A. Thorpe, *Lecture Notes on Elementary Topology and Geometry*. Springer-Verlag, New York (1967).

B.J. Skinner and S.C. Porter, *The Dynamic Earth*. Wiley, New York (1992).

B. Spain, *Tensor Calculus, A Concise Course*. Dover, New York (2003).

H. Štefančić, Expansion around the vacuum equation of state: sudden future singularities and asymptotic behavior, *Physical Review D* **71**, 084024 (2005).

D.R. Stevenson and C.J. Thompson, Fairy ring kinetics, *Journal of Theoretical Biology* **58**, 143–163 (1976).

S. Svanberg, *Atomic and Molecular Spectroscopy, Basic Aspects and Practical Applications*, second edition. Springer-Verlag, New York (1992).

C. Swithinbank, The origin of dirt cones on glaciers, *Journal of Glaciology* **1**, 461–465 (1950).

T. Tacheuchi, *An Illustrated Guide to Relativity*. Cambridge University Press, Cambridge, UK (2010).

J.L. Tonry *et al.*, Cosmological results from high-z supernovae, *Astrophysical Journal* **594**, 1–24 (2003).

P.B. Topham, Colonization, growth, succession and competition, in M.R.D. Seaward (ed.), *Lichen Ecology*, pp. 31–68. Academic Press, London (1977).

V.C. Tsai1 and G. Ekstrom, Analysis of glacial earthquakes, *Journal of Geophysical Research* **112**, F03S22/1–13 (2007).

V.C. Tsai, J.R. Rice, and M. Fahnestock, Possible mechanisms for glacial earthquakes, *Journal of Geophysical Research* **113**, F03014/1–17 (2008).

A. Turing, The chemical basis of morphogenesis, *Philosophical Transactions of the Royal Society B* **237**, 37–72 (1952). Reprinted in *Bulletin of Mathematical Biology* **52**, 153–197 (1990).

M.F. Twight, *Extreme Alpinism, Climbing Light, Fast, and High*. The Mountaineers, Seattle (1999).

J. Tyndall, On the blue colour of the sky, the polarization of skylight, and on the polarization of light by cloudy matter generally, *Proceedings of the Royal Society of London* **17**, 223–233 (1868–1869).

J. Tyndall, *Hours of Exercise in the Alps*. D. Appleton & C., New York (1872).

C.J. van der Veen, *Fundamentals of Glacier Dynamics*, second edition. CRC Press/ Taylor & Francis, Boca Raton, FL (2013).

Very Large Telescope (VLT) website http://www.eso.org/public/teles-instr/paranal-observatory/vlt/

R.M. Wald, *General Relativity*. Chicago University Press, Chicago (1984).

H.J. Weber and G.C. Arfken, *Essential Mathematical Methods for Physicists*. Elsevier-Academic Press, Amsterdam (2004).

S. Weinberg, *Gravitation and Cosmology*. Wiley, New York (1972).

J. Weiss and B. Zumino, Supergauge transformations in four dimensions, *Nuclear Physics B* **70**, 39–50 (1974).

J. Weiss and B. Zumino, A lagrangian model invariant under supergauge transformations, *Physics Letters B* **43**, 52–54 (1974).

V.F. Weisskopf, Of atoms, mountains, and stars: a study in qualitative physics, *Science* **187**, 605–612 (1975).

V.F. Weisskopf, Search for simplicity: mountains, water waves, and leaky ceilings, *American Journal of Physics* **54**, 110 (1986).

V.F. Weisskopf, Physics of mountains — response, *American Journal of Physics* **54**, 871 (1986).

J.A. Wheeler and K. Ford, *Geons, Black Holes, and Quantum Foam: A Life in Physics*. W.W. Norton & C., New York (1995).

D. Worster, *The Life of John Muir*. Oxford University Press, Oxford, UK (2008).

S.H. Yang and Y.L. Shi, Three-dimensional numerical simulation of glacial trough forming process, *Science China (Earth Sciences)* **58**, 1656–1668 (2015).

M.W. Zemansky and R.H. Dittman, *Heat and Thermodynamics*, seventh edition. McGraw-Hill, New York (1997).

Index

www.ingramcontent.com/pod-product-compliance
Lightning Source LLC
Chambersburg PA
CBHW050553190326
41458CB00007B/2024